The *Springer Series in Advanced Microelectronics* provides systematic information on all the topics relevant for the design, processing, and manufacturing of microelectronic devices. The books, each prepared by leading researchers or engineers in their fields, cover the basic and advanced aspects of topics such as wafer processing, materials, device design, device technologies, circuit design, VLSI implementation, and sub-system technology. The series forms a bridge between physics and engineering, therefore the volumes will appeal to practicing engineers as well as research scientists.

More information about this series at http://www.springer.com/series/4076

Klevis Ylli · Yiannos Manoli

Energy Harvesting for Wearable Sensor Systems

Inductive Architectures for the Swing Excitation of the Leg

Springer

Klevis Ylli
Hahn-Schickard
Villingen-Schwenningen,
Baden-Württemberg, Germany

Yiannos Manoli
Department of Microsystems Engineering
University of Freiburg
Freiburg, Baden-Württemberg, Germany

ISSN 1437-0387 ISSN 2197-6643 (electronic)
Springer Series in Advanced Microelectronics
ISBN 978-981-33-4447-1 ISBN 978-981-33-4448-8 (eBook)
https://doi.org/10.1007/978-981-33-4448-8

This Springer imprint is published by the registered company Springer Nature Singapore Pte Ltd.
The registered company address is: 152 Beach Road, #21-01/04 Gateway East, Singapore 189721, Singapore

Preface

The term "Energy Harvesting" describes the generation of small amounts of electrical energy from the surrounding environment, be it kinetic energy, light or heat gradients. This technology offers the potential for the low-maintenance operation of wireless sensor systems without the need to charge or replace batteries. In order to power electronic systems on the human body, harvesting devices are being developed that use the body as an energy source. This book focuses on the human gait as a source of kinetic energy. The swing motion of the foot during walking provides the necessary excitation for flat inductive devices that can be integrated into the shoe sole. These devices are, in principle, comparatively simple to understand and fabricate. The main difficulty lies with the optimization of the geometrical parameters.

Several harvester architectures are investigated in this book, in order to find the magnet-coil arrangement that generates the largest power output. Before devices can be fabricated, the optimal geometrical parameters for the highly restricted space in the shoe sole need to be determined. A system model based on a differential equation of motion is developed that takes the relevant physical effects into account, i.e. the forces that act on the moving magnets and affect their motion. The steps taken to implement the required aspects of the system model in a numerical simulation environment are described in detail, and the model is subsequently used to calculate the motion of the magnets relative to the coils depending on the external excitation. The required simplifications as well as the drawbacks and limitations of the system model are described.

Based on the magnet motion, the induced voltage and the generated power output are determined. While the magnetic field distribution needs to be calculated for every setup separately, the system model is flexible and used across all architectures. The geometrical parameters of each architecture are variable, and an optimization algorithm is used to maximize the power output for each architecture.

The fabrication of harvester devices as closely as possible to the optimized design is the next step. The devices are characterized on a treadmill with two different test persons and at walking speeds of 4, 6, 8 and 10 km/h. The power output is determined using an ohmic load equal to the coil resistance. Of the four analyzed architectures, the highest achieved average power output is 10.32 mW. In terms of power density, this translates into 1.15 mW/cm^3 for an active volume of 8.96 cm^3. The maximum device dimensions including the housing remain the same throughout this work at 77 mm in length, 41.5 mm in width and 15.75 mm in height (resulting in a total volume of 50.33 cm^3).

The experimental results are used to revise and verify the system model. A new parameter optimization is performed for each architecture using the revised model. The second generation of fabricated devices (device volume remains 50cm^3) achieves a greatly improved average power output of up to 43 mW during walking (power density of 0.85 mW/cm^3).

Experiments are also carried out in industrial applications on a pneumatic piston and a pneumatic clutch-brake with power outputs reaching up to 15 mW. These results prove that this type of linear, non-resonant device can also be used in industrial applications where the excitation, while certainly not identical to human motion, shows a few similarities like non-resonant behavior and large amplitudes.

Lörrach, Germany
Freiburg, Germany

Klevis Ylli
Yiannos Manoli

Contents

Nomenclature

Abbreviations

BLE	Bluetooth Low-Energy
BWS	Body-worn systems
CMS	Condition-monitoring system
DOE	Design of experiments
edf	Electrical damping force
EH	Energy harvester
FEM	Finite element method
GCS	Global coordinate system
GP	Geometrical parameters
HAC	Harvester architecture
INT1	Integrator 1 of the system model, etc.
IR	Infrared
LCS	Local coordinate system
LED	Light-emitting diode
MOGA-2	Multi-Objective Genetic Algorithm, used in parameter optimization
N35	Polarization strength of NdFeB permanent magnets
NdFeB	Neodymium iron boron (rare earth permanent magnet material)
O1	Data set from first optimization run. O2 Analogously
PCB	Printed circuit board
PV	Photovoltaic (cell)
RF	Radio-frequency
RMS	Root-mean-squared value
S1	Piston setup S1 in laboratory experiments. S2 and S3 analogously
WSN	Wireless sensor node

Symbols

A	Area enclosed by coil windings (m^2)
A_{ext}	External acceleration (e.g. from measurements of the foot motion) (m/s^2)
b_{mag}	Thickness of rectangular magnets (m)
B	Magnetic flux density (T)
B_{mean}	Mean flux density (e.g. for a given coil winding) (T)
B_x	X-component of magnetic flux density B (perpendicular to the direction of motion of the magnets in a cuboid setup) (T)
B_z	Z-component of magnetic flux density B (along the direction of motion of the magnets) (T)
c_v	Transition approximation coefficient of the friction model (s/m)
distInnen	Vector of the coil positions within the harvester housing. Used in the system model
d_{coil}	Thickness of cylindrical and rectangular coils (m)
d_{Cu}	Diameter of the copper wire used to wind the coils (m)
d_{iron}	Back flux guide thickness (“back iron”) (m)
d_{mag}	Wall thickness of ring magnets (m)
d_{Spacer}	Thickness of steel spacers (m)
f_c	Viscous friction coefficient (N/m/s)
F_{brk}	Breakaway friction force (N)
$F_{brk,S}$	Static breakaway friction force (N)
F_C	Coulomb friction force (N)
F_{el}	Electrical damping force (N)
F_R	Friction force (N)
F_S	Spring force (N)
F_T	Inertial force (N)
F_V	Viscous friction force (N)
g	Earth’s gravity (m/s^2)
h_{coil}	Length of the edge of the rectangular coil located along the direction of motion (m)
h_{core}	Size of the air gap in the middle of the coil (same direction as h_{coil}) (m)
h_{mag}	Magnet height (m)
H_c	Coercive field strength of magnets (A/m)
$k(x)$	Electro-magnetic coupling function (V/m/s)
k_i	Coefficient of restitution (hardstop model) (–)
l_{coil}	Length of the edge of the rectangular coil located perpendicular to the direction of motion (into the drawing plane of the FEM-simulation). Coil height in the case of cylindrical coils (m)
μ_k	Coefficient of kinetic friction (–)
μ_s	Coefficient of static friction (–)
μ_{mag}	Relative permeability (–)

m	Number of coil windings in the direction perpendicular to the winding area of a coil (–)
m_{mag}	Magnet stack mass (including steel flux guides where applicable) (kg)
n	Number of coil windings in the radial direction of a cylindrical coil, or in the direction of motion in the case of rectangular coils (–)
n_{coils}	Number of coils simulated/used in a harvester (–)
n_{mag}	Number of magnets simulated/used in a harvester (–)
N	Number of total windings of a coil (–)
φ	Magnetic flux (Wb)
P_{avg}	Average power output (mW)
P_N	Normalized power output (–)
ρ	Specific resistance of copper wire (Ω/m)
r_i	Radius of coil winding i (m)
r_{mag}	Inner magnet radius in the case of cylindrical magnets (m)
r_{mean}	Mean winding radius of a cylindrical coil (m)
R_{in}	Resistance of a coil (Ω)
R_l	Ohmic load resistance (Ω)
R_{mag}	Outer magnet radius in the case of cylindrical magnets (m)
R_{max}	Radius of the outermost (largest) winding of a cylindrical coil (m)
R_{min}	Radius of the innermost (smallest) winding of a cylindrical coil (m)
u_i	Logic signal number i in the Simulink environment (–)
U_{ind}	Induced voltage (V)
U_l	Voltage across the ohmic load (V)
v_{thr}	Threshold speed of the magnet in the friction model (m/s)
$\ddot{x}$	Magnet acceleration (m/s^2)
$\dot{x}$	Magnet speed (m/s)
x	Magnet position within housing (m)
x_{max}	Total motion space within the harvester housing (m)

List of Figures

List of Tables

Chapter 1
Introduction

In today's world, technology is leaping ahead every day. New developments in recent years have lead to the highly increased use of portable devices. These enable the user to communicate and interconnect with other people, to make use of integrated sensors and to retrieve information via wireless connections on the go without the limitations of cable-based devices. Body-worn systems (BWS) are also gaining importance and a wide range of companies is bringing their devices to the market. Body-worn systems exist in various forms, from a simple fitness tracker that counts your steps during walking, up to a complex bio-medical monitoring system. These devices assist the user, facilitate daily life and in some cases can even extend a persons freedom, e.g. by being medically monitored outside of their house. As is the case with other portable devices, the aim is to integrate these devices into everyday life and daily routines as unobtrusively as possible.

While many portable devices have reached a certain maturity, with smartphones being the most prominent example, there are still numerous issues that need to be overcome. The size and weight of a portable device is an important factor concerning the comfort perceived by the user. The most limiting factor across nearly all device types however is the battery lifetime. Modern devices contain powerful electronics that can fulfill a multitude of tasks at the cost of a comparatively high power consumption. The developments in low-power systems and low-power RF technologies are leading to a reduction in the overall power consumption. Nevertheless, the users of these products have grown accustomed to having to charge several devices several times a week. This inconvenience is a major annoyance and cannot easily be avoided.

An approach to solve the power-supply issues for mobile low-power systems is to use energy harvesting (EH). The term "energy harvesting" describes the generation of small amounts of electrical energy from the energy that is available in the surrounding environment. Some ideas from large-scale renewable energy applications can be adapted to the small scales of energy harvesting, e.g. using fluid flow or light as an

K. Ylli and Y. Manoli, *Energy Harvesting for Wearable Sensor Systems*,
Springer Series in Advanced Microelectronics 62,
https://doi.org/10.1007/978-981-33-4448-8_1

energy source, while several other methods have been developed specifically for the usually very compact energy harvesting devices. The topic of energy harvesting has been in the focus of research groups globally for approximately two decades. Energy harvesting offers a maintenance-free solution for low-power systems that makes the tedious and costly process of battery replacements or charging obsolete.

1.1 Literature Review

In many cases, the EH devices are developed to generate energy in industrial environments, e.g. from the vibrations of motors, gearboxes and pump. Condition-monitoring systems are an exemplary application that requires a power supply [1]. Particularly in the case of refurbished systems, it is often not possible to change the wiring in order to provide power to a sensor node. The usually completely wireless sensor nodes (WSN) are required to be autonomous systems, which contain a sensor unit, a processing unit, a wireless transmission technology as well as an autonomous power supply.

Apart from the industrial environment, a significant amount of research is dedicated to the human body as an energy source. Possible energy sources include the body heat [2], light captured by photovoltaic (PV) cells on the body surface and the body motion. Kinetic energy is abundantly available during human everyday activities, particularly during walking [3, 4]. The power expended by the body to perform certain motions can reach several dozen Watts [3]. An overview of the state of the art and different energy harvesting methods for kinetic harvesting on the human body and shoe-based devices in particular, is given in the following sections.

1.1.1 Kinetic Energy Harvesting from the Human Gait

It was mentioned before, that the body uses the equivalent of dozens of Watts of power for motions performed during walking. This energy can partially be harvested using kinetic energy harvesters. Amongst the body locations, where harvesting devices have been applied in literature are the wrist, the back (by using backpacks), the knee, the ankle and the shoe. The shoe or foot performs a similar motion to the ankle during walking. It offers some space for the unobtrusive integration of a harvesting device in the shoe sole and is thus limited by other boundary conditions than devices placed at the ankle. Shoe-based devices are therefore considered separately in the following sections.

Harvesters for placement at the wrist are most commonly designed as a pendulum structure. In one approach, the pendulum mass consists of a magnet setup and moves past coils to generate energy as the arm moves during walking [5]. In the approach shown in Fig. 1.1a, the method of magnetic plucking is used. A pendulum mass that contains a magnet can rotate freely within the housing. During each rotation the pendulum magnet moves past several radial beams made of piezoelectric material with magnets attached at their tips. The magnetic forces deflect the piezo beams and they can vibrate freely at their Eigenfrequency when the force is released, as the pendulum rotates away. When the piezoelectric beams are deflected, charges are separated in the beam due to asymmetries in the crystal lattice and effectively generate a voltage between two sides of the beam, where electrodes are placed to allow further processing of the voltage. The power generation reported for this device is on the order of a few micro-Watts [6].

Backpack harvesting devices use a proof mass within the backpack which can oscillate vertically following the low-amplitude sinusoidal motion of the hips and upper body during walking. The proof mass may consist of magnets that move past coils, or it can directly drive a gearbox and a generator through its motion [7]. The gearbox translates the small motion of the proof mass into a large rotational velocity of the generator to achieve power outputs of up to 9 W. However, a large proof mass is required to reach such a high power output, with backpack loads of up to several dozen kilograms being used in literature [8–10]. Another approach uses shoulder straps made of piezoelectric material which generates a voltage as it is stretched due to the backpack load [9].

Knee-based energy harvesting devices rely on the relative motion between the lower and upper leg (rotation at the knee) to harvest energy. There are approaches with piezoelectric materials that bend at the knee to generate energy [11], or that use magnetic plucking of piezoelectric beams [12]. These devices generate an electrical power output of a few micro-Watts [11] or a few dozen milli-Watts [12]. Other devices use a gearbox to translate the small rotational motion at the knee into a high number of revolutions driving an electromagnetic generator [4]. These devices are capable of generating up to 6 W of electrical power [4].

Devices presented in literature for a placement at the ankle are of a linear inductive type as shown in [13] or of a pendulum inductive type [14] as shown in Fig. 1.1b. This device contains an eccentric tungsten proof mass which drives a magnetic wheel divided into eight magnet segments of 45° each, as it rotates. The magnets move past eight coils integrated in the housing to generate an average power output of approximately 30 mW across an ohmic load.

1.1.2 Examples of Piezoelectric Harvesting Devices

One of the earliest implementations of piezoelectric energy harvesting in a shoe sole was performed two decades ago [15]. Therein a "TH 7R" PZT unimorph (with an area of roughly 49 cm^2) and an eight-layer stack of 28 μm thick PVDF sheets (with an area of roughly 65 cm^2) were placed within a shoe (between the rubber sole and the insole) to generate average power outputs of 1.8 mW and 1.1 mW respectively during walking.

In the years since, a wide range of structures has been presented, that harvests energy from the walking motion using the piezoelectric effect. A few of these most

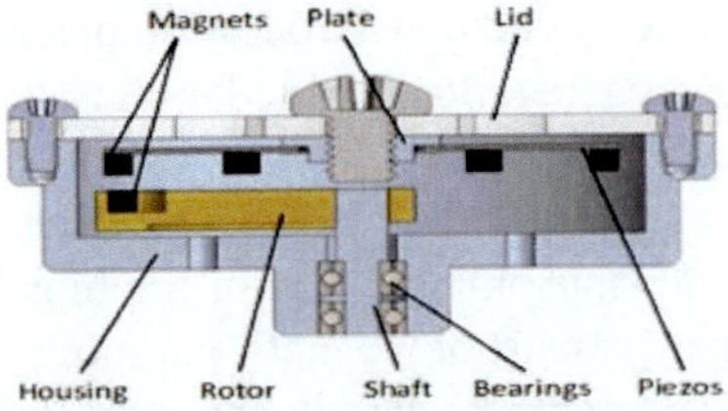

(a) Wrist Harvester [6]

(b) Ankle Harvester [14]

Fig. 1.1 Exemplary kinetic harvesting devices for various locations on the human body

recurring structures are listed below. Several papers focus on the weight of the user to deform the piezoelectric material as shown above [16, 17]. Other approaches for example use the heel strike to excite piezoelectric cantilever structures that then vibrate freely at their resonance frequency [18]. A lot of effort is also directed towards researching functional nanofibers, with the idea being, that in the future a woven piece of high-tech-fabric may contain hundreds or thousands of piezoelectric fibers [19]. While the piezoelectric effect is quite common amongst harvesting devices and well-understood, the achieved power outputs in the above examples reach a few mW at best. The next section shows two capacitive examples that promise a superior power output.

1.1.3 Examples of Capacitive Harvesting Devices

Two capacitive principles of harvesting energy from human motion shall be mentioned here, as they offer the potential for very high power outputs. The first principle uses dielectric elastomers, a polymer material which expands when a voltage is applied across it [20] and can undergo large deformations [21]. Two flexible electrodes on either side of the polymer film (elastomer generator) effectively turn it into a variable capacitor. As the material is stretched mechanically, the capacitance increases. A bias voltage is applied, then the mechanical strain is removed (e.g. through heel lift-off) and the polymer returns to its original shape. The capacitance thus decreases and electrical energy is produced. An early bellow-like structure [20] implemented in a shoe is shown in Fig. 1.2. The structure is used to allow the elastomer to expand through the holes and increase its area upon heel strike. This device was able to generate up to 800 mJ per cycle, thus reaching average power outputs of nearly one Watt. The drawbacks of this technique are the high voltages that

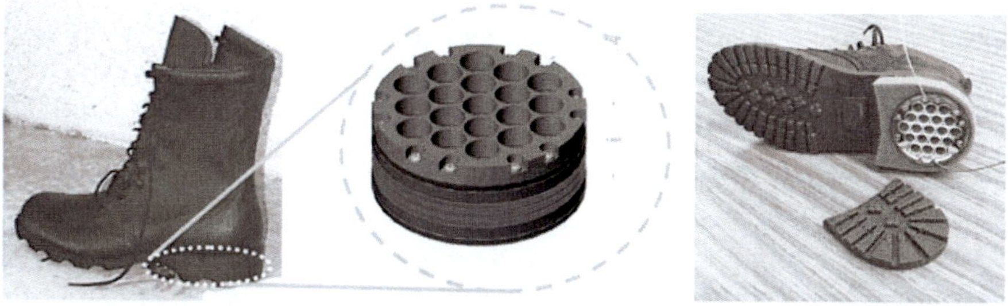

Fig. 1.2 The experimental shoe from [20] is shown, with a bellow-like structure underneath the heel, which is used to expand the dielectric elastomer upon heel strike. Figure from [20]

are required for an efficient operation of the dielectric elastomer generator. While research on this topic has continued as shown by [21] and other publications, no products employing this principle have been presented on the market.

The second harvesting principle is called REWOD (reverse electro-wetting on dielectric) and in simple terms, it describes a variable capacitor, where the change in capacitance is caused by liquid components, i.e. either through changing of liquid dielectrics with liquid conductors between the capacitor plates [22], or by making one of the "capacitor plates" a liquid itself [23].

The latter variant's schematic from [23] is shown in Fig. 1.3. The device consists of a silicon substrate (green in Fig. 1.3) which has holes in it. A metal electrode with a layer of insulation on top is placed around each whole (red in the schematic). The top electrode is a conductive liquid (mercury) which covers the whole volume between the silicon substrate and the top part of the device which consist of a membrane that does not let the liquid through (yellow). It does, however, let air bubbles through. The idea is that by forcing air through holes from the bottom to the top of the device, the conducting liquid is pushed away from the bottom electrode with each air bubble.

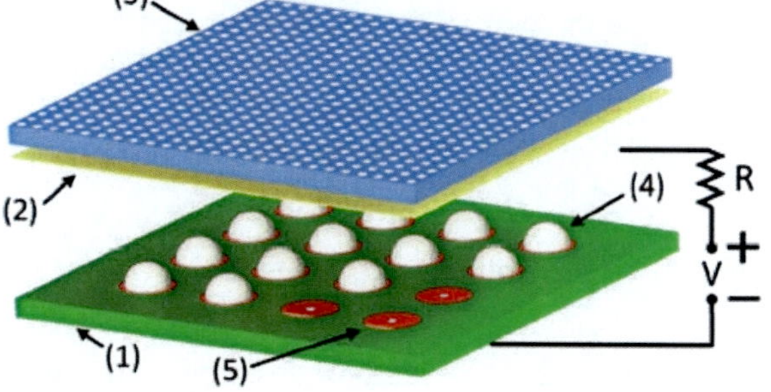

Fig. 1.3 Schematic of the "Bubbler" REWOD device. The fixed electrodes covered in insulating material (5) are shown in red on the green silicon substrate (1). The white balls are the air bubbles (4). (2) and (3) are the membrane and top plate through which the air bubbles can escape. Figure from [23]

By designing the device dimensions accordingly, air bubbles can be made to grow to a certain size before touching the top membrane and escaping from the capacitor. As a bubble escapes, the liquid moves back onto the electrode, effectively reducing the distance between the two capacitor plates. In simulations, the holes were varied with diameters ranging from 140 μm to 200 μm and the gap between the electrodes and the top membrane was varied between 100 μm and 130 μm to find the best geometry. The designed 40 mm × 40 mm REWOD chip consists of 800 electrodes of 1 mm diameter. With a given pressure during walking, the air bubbles are expected to be generated at about 2 kHz which using a bias voltage of 30 V sums up to an expected average power output of 1 W. This is certainly one of the most promising approaches today.

1.1.4 Examples of Triboelectric Harvesting Devices

Triboelectric energy harvesters are based on the cycling contact and separation of two different materials. As the two materials come into contact, a flow of electrons from one material to the other occurs due to opposite triboelectric polarities [24]. The device from [24] is shown in Fig. 1.4. It consists of a folding Kapton substrate (shown in yellow color in the schematic) onto which alternating layers of PTFE (polytetrafluoro-ethylene) and aluminum have been applied (depicted in green and blue color respectively). As weight is applied during walking, the folding structure is compressed and the aluminum and PTFE layers come into contact. The instantaneous short-circuit current was shown to reach up to 0.66 mA while the open-circuit voltage reached up to 215 V for the shown experimental structure with a total size of 3.8 cm × 3.8 cm × 0.95 cm. With a load of 1 MΩ a maximum instantaneous power output of 4.2 mW was achieved by compressing the structure by hand (up to roughly 400 N). No information was given on average power outputs or the power output during walking.

A second, similar example [25] is given in Fig. 1.5. Therein a multilayered structure is used to combine the piezoelectric and triboelectric effect within a compact and flexible generator of 2.5 cm × 2.5 cm. A short-circuit current of 25 μA and an open-circuit voltage of 280 V were achieved.

1.1.5 Examples of Inductive Harvesting Devices

In literature a range of inductive devices has been presented, which harvest energy from the human gait. One subset of these devices uses the heel strike, i.e. the contact between the heel and the ground, as an excitation source. These are resonant devices where a spring-loaded inertial mass, usually a magnet arrangement, is deflected upon heel-strike and can vibrate freely at its resonant frequency [26]. This operation principle is named frequency up-conversion as the slow motion of the legs with a

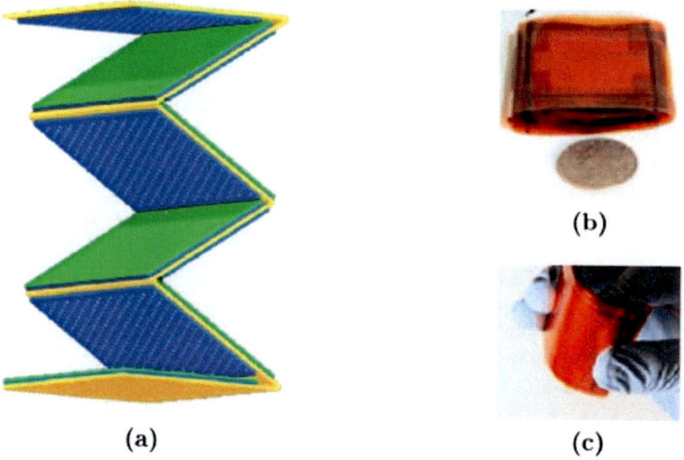

Fig. 1.4 Schematic of the folding-structure triboelectric generator in (**a**). The Kapton substrate is shown in yellow, while the alternating layers of PTFE and aluminum are shown in green and blue respectively. The fabricated device with total dimensions of 3.8 cm × 3.8 cm × 0.95 cm is shown in (**b**) and (**c**). Figures from [24]

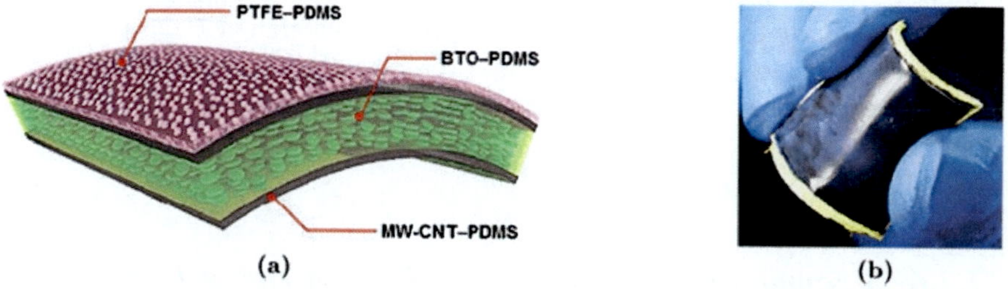

Fig. 1.5 Schematic of the multilayered device shown in (**a**). The piezoelectric material PDMS (polydimethylsiloxane) is embedded within different materials to enable the triboelectric effect. The fabricated device is shown in (**b**). Figures from [25]

step frequency of 1 Hz is used to set an inertial mass into motion at its own, far higher resonant frequency. A device that uses the heel strike and is designed in a flat manner to fit within a shoe is shown in Fig. 1.6a. A magnetic circuit that moves past a fixed coil when excited is suspended magnetically using a spring magnet in the bottom and top of the housing. The generated voltage curve is shown in Fig. 1.6b. As these devices are excited vertically, the available motion amplitude within the shoe sole is very small. Apart from inductive devices, there is also a range of piezoelectric devices based upon the acceleration pulse upon heel strike [27].

Several harvesting devices have been presented that are not prepared for integration into the shoe. They use hand-vibrations [28–30], or the heel strike while attached at the ankle [13]. These resonant devices use magnet stacks that oscillate within coils to generate energy. Still, they are worthwhile mentioning in this section, because the working principle is very similar to devices that use the swing motion of the leg

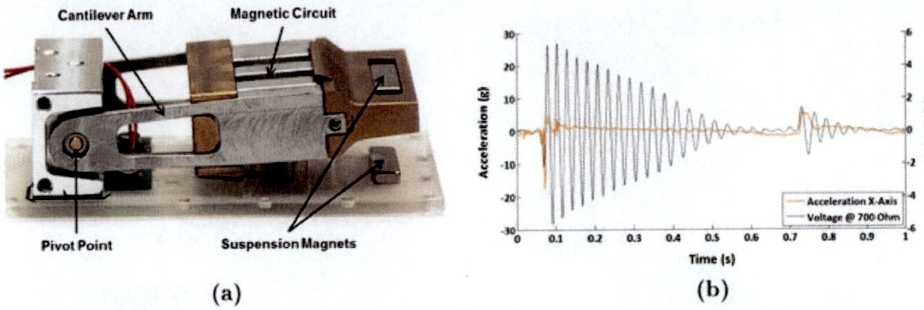

Fig. 1.6 Exemplary frequency up-conversion heel strike device in (**a**) and the generated voltage curve with a decaying amplitude in (**b**). Figures from [26]

during walking as excitation source. Devices [13, 28–30] could be adapted in the future, to generate energy from the leg swing, as do the devices presented in this book.

Amongst the earliest to try and fit a generator to the shoe was MIT's Media Laboratory with two rotary electromagnetic generators presented in [15, 31]. They are based on the mechanism of hand-cranked flashlights and the heel strike of the user is applied to the generator using a gearbox with a high gear ratio. While these devices are able to generate average power outputs of up to 250 mW, they are also very large and heavy, complex in terms of their mechanics and contain parts outside of the shoe.

1.1.6 Examples of Inductive Swing-Excited Harvesting Devices

Devices that use the swing motion of the leg as an excitation source are either set up as non-resonant devices without any springs or as devices with non-linear magnetic springs which exert no force on the moving magnets, if the distance between moving magnet and spring magnet is large enough. This is possible because of the exponential decay of the repelling force between magnets with the distance between them.

Figure 1.7 shows such a device that uses a magnetic slider within a tube as well as magnetic springs at either end of the tube [13]. The device setup schematic is given in Fig. 1.7a and the fabricated device is shown in Fig. 1.7b. The device is fabricated with two different total lengths of 140 mm and 200 mm and two different magnet stacks consisting of either three or six magnets. Experiments with the device attached on the lower leg near the ankle to make use of the leg's swing motion, show that the long device with a stack of three magnets generates by far the most power of the analyzed setups. At a walking speed of 6 km/h the device generates an average power of 6.6 mW. This was measured with an ohmic load of 5 Ω, which is equal to the coil resistance of the device. The experimental outcome can be attributed to the distance

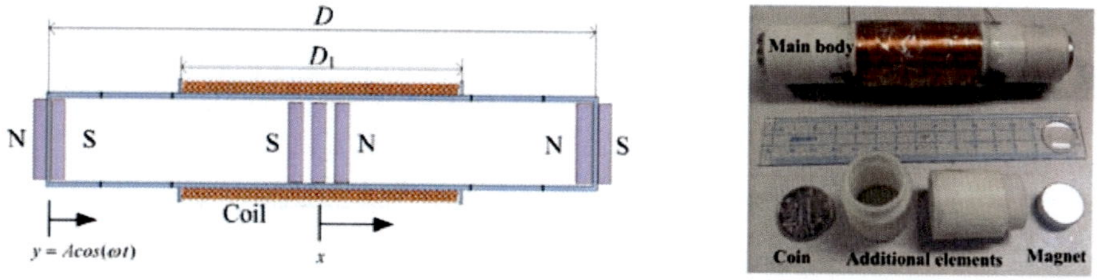

(a) (b)

Fig. 1.7 First exemplary swing-excited energy harvester with the device schematic shown in (**a**) and the fabricated device shown in (**b**). Figure from [13]

between the moving magnet stack and the spring magnets. In a shorter device, the spring magnets are closer to the moving magnet and consequently the repelling force has a larger effect on the magnet motion.

It is also shown, that the device produces more power with increasing walking speed. Using this devices in a vertical orientation on the lower leg to make use of the heel strike was also analyzed experimentally and showed a far smaller power output than in the horizontal setup (approx. 2 mW at 6 km/h walking speed). However, at the given length and a diameter of 25 mm, the device is generally far too large for the integration into a shoe.

A second swing-excited device is presented in [32]. It uses a magnet stack made of ring magnets in opposing polarization with flux guides between them. The fabricated magnet stack consists of a total of 16 magnets and 17 flux guides. The coils (30 in total) are attached directly to the plastic channel through which the magnets can move. From the mentioned size of the components the total device is estimated to be approximately 120 mm in length and 20 mm in diameter. As was the case with the previous device, this device was not integrated into a shoe and does not meet the size constraints for integration that are introduced in Sect. 1.3 of this work.

The device is divided into three phases with 10 coils per phase and a resistance of 19.8 Ω per phase. The power output is not given explicitly, however it is mentioned that during testing the load current was 1.5 mA with a 1000 Ω load resistor. This would translate into an average power output of 2.3 mW. The walking speed of the experiment is not mentioned.

Finally a third swing-excited device [33] is shown in Fig. 1.8. Its schematic is depicted in Fig. 1.8a and as can be noted from Fig. 1.8b, it is the only swing-excited device that was integrated into a shoe. In this study again the cylindrical magnet-within-coil setup is investigated. The device size is fixed to a length of 100 mm and a diameter of 15 mm. The diameter is within the size constraints that are defined in Sect. 1.3, however it has to be kept in mind, that the device does not have a protective housing. The length is noticeably larger than for the devices presented in this book.

Several magnet dimensions and coil dimensions were investigated for the given device size to optimize the power output. The coils are either regarded separately with a separate load resistor for each coil or in a series connection. When considering

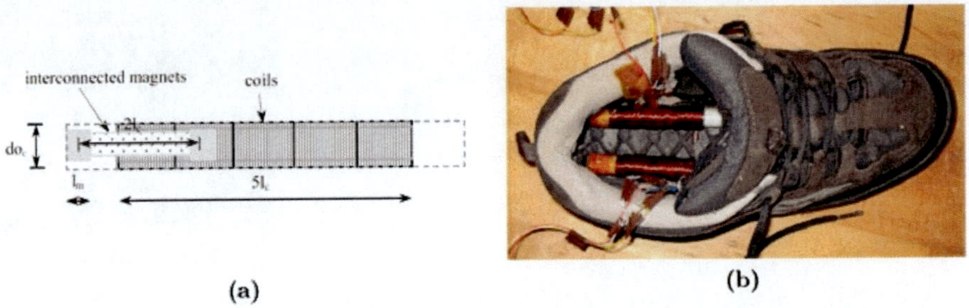

Fig. 1.8 Third exemplary swing-excited energy harvester with the device schematic shown in (**a**) and the fabricated device shown in (**b**). This is the only device that was actually integrated in a shoe (albeit without any particular protection of the transducer). Figure from [33]

Table 1.1 Selection of parameters from swing-excited inductive harvesters presented in literature. Masses and volumes are estimated, when not specifically given in the sources. Table from [34]

P_{avg}	Excitation	Magnet stack mass	Volume [cm^3]	Power density [mW/cm^3]	Authors
6.6 mW	6 km/h walk	94 g	98.17	0.067	Wang et al. [13]
2.3 mW	unknown	43.7 g	37.70	0.061	Shen et al. [32]
7.9 mW	6 km/h walk	12.4 g	17.67	0.447	Caroll et al. [33]

single coils separately, the largest power output is achieved with a single magnet and three coils at nearly 14 mW (10 Ω load resistance). The power output for a series connection of three coils is not given, however, for four coils it is 7.9 mW when two magnets are used and approximately 5 mW when a single magnet is used. The walking speed used in the experiment is mentioned as "2 steps/second" [33], because of two distinct bundles of oscillating voltage peaks in the generator signal. From the experience obtained during this work, the two bundles of voltage peaks are generated during lift-off and as the foot decelerates before touching down on the ground. The two voltage bundles thus amount to one step per second. In the measurements performed for this work, it is shown that a step frequency 1 Hz occurs at a walking speed of approximately 6 km/h (see Sect. 1.2 for an analysis of the human gait).

Table 1.1 from [34] summarizes the main parameters of the above devices. In this work the coils are always connected with each other in order to produce a single alternating generator signal that can be rectified and processed. Therefore the power output of 7.9 mW from [33] is used as the reference in Table 1.1.

Of the devices presented in literature that use the swing motion of the leg as an excitation source, only the device from Carroll et al. [33] has actually been integrated in the shoe. Additionally, the focus has mainly been on cylindrical structures as far as the swing motion of the leg is concerned. This work investigates several cylindrical and rectangular architectures for direct integration within the constraints of the shoe sole as is shown in the following sections.

1.2 Human Gait as an Excitation Source

The human gait exhibits three possible sources of energy for energy harvesting devices, which are schematically depicted in Fig. 1.9. There are two acceleration-based excitation sources, namely the acceleration pulse upon heel strike, i.e. the impact between shoe and ground, and the acceleration due to the leg swing during walking. The other excitation source is the force that acts upon the shoe due to the weight of the person. As far as the force acting upon the shoe sole is concerned, previous investigations have shown that the force can dynamically amount to up to 130% of the person's body weight [3] and therefore easily reach values of 1000 N. The force develops in the shape of the letter "M" [35] as it shifts from acting fully on the heel upon heel strike, towards the forefoot upon lift-off.

In this work, the swing motion of the leg is considered as a potential energy source for harvesting energy from the human gait. In order to obtain acceleration data, acceleration measurements were performed at constant motion speeds of 4 km/h, 6 km/h, 8 km/h and 10 km/h on a treadmill. A three-axis accelerometer was firmly attached to the sole at the back of a shoe. For a shoe lying flat on the floor, the x-axis was in line with the horizontal axis of the shoe (pointing forwards). The accelerations were recorded with a duration of one minute for each motion speed. Two different runners with different body geometries were recorded with runner "2" being lighter and taller (65 kg, 1.85 m), while runner "1" is heavier and smaller (110 kg, 1.8 m). The recorded acceleration data is the basis for the system simulations in the following chapters and is used as an input into the system model. However, it must be noted, that the data of the two test persons is of limited statistical significance as it represents the unique walking style and particular body geometry of those two persons.

1.2.1 Accelerations of the Foot During Walking

A distinctive feature of walking is the very low frequency at which successive footsteps occur. Table 1.2 gives the step frequencies for two test persons at various motion speeds. The step frequency of one foot varies between 0.8 Hz and 1.2 Hz as a function of motion speed. It can be seen that there is only a moderate increase in the step frequency with increasing motion speed. Towards higher motion speed the increase in step frequency diminishes, implying that it is the step length which has to increase to allow faster motion.

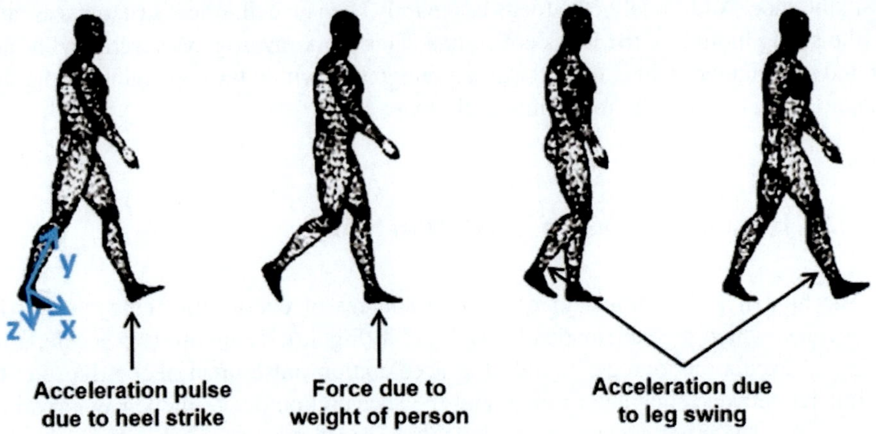

Fig. 1.9 Possible acceleration sources of the human gait. The measurement axes of the acceleration sensor are shown in blue color. The x-axis points forward along the shoe and the y-axis upward along the leg. The z-axis points away sideways from the body and is of no relevance in the human gait

Table 1.2 Step frequencies for two test persons (obtained from treadmill acceleration measurements). Table from [34]

Motion speed		Step frequency: Person "1"	Step frequency: Person "2"
4 km/h	slow walking	0.82 Hz	0.8 Hz
6 km/h	fast walking	1.0 Hz	0.94 Hz
8 km/h	slow running	1.18 Hz	1.2 Hz
10 km/h	fast running	1.19 Hz	1.19 Hz

Figure 1.10 shows 2.5 s time frames of the acceleration data along the x-axis of the shoe. The data shown was recorded at 4 km/h and 10 km/h, thus giving an overview of the acceleration patterns for both walking and jogging. One observation made for the acceleration data is the large value of the acceleration peaks. Vertical accelerations on the y-axis can peak at nearly 50g while on the horizontal axis they can peak at approximately 20g. Table 1.3 summarizes the average acceleration peaks for the full data recordings for the x and y axis and two different runners respectively. The term "average accelerations peaks" is explained by the green and red markers in Fig. 1.10. It means that the peaks in the acceleration data are found by a peak detection algorithm and the average value of all the detected peaks gives the results in Table 1.3 from [36]. On the x-axis considered herein, the positive values correspond to a forward motion of the foot. The separate phases of the human gait are highlighted in Fig. 1.10.

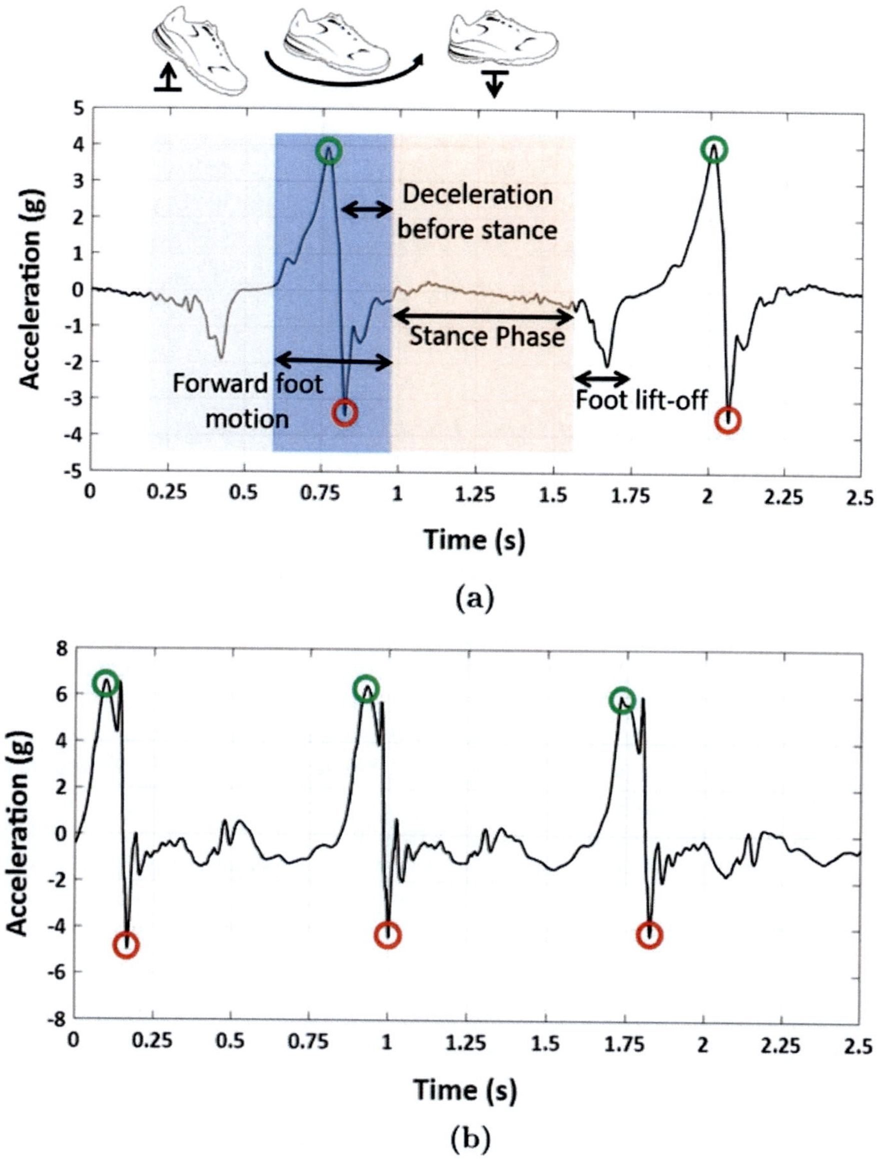

Fig. 1.10 Accelerations recorded at the shoe sole on the x axis at a motion speed of 4 km/h in (**a**) and 10 km/h in (**b**) for test person "1". The green and red markers indicate acceleration peaks found by a peak detection algorithm. The separate phases of the human gait are highlighted in (**a**)

Table 1.3 Average acceleration peaks recorded on a treadmill with two test persons under different walking speeds. Table from [36]

Motion speed	Positive peak average [g]				Negative peak average [g]			
	Person "1"		Person "2"		Person "1"		Person "2"	
	x-axis	y-axis	x-axis	y-axis	x-axis	y-axis	x-axis	y-axis
4 km/h	4.4	5.2	2.9	6.4	−3.7	−10.2	−4.9	−10.5
6 km/h	5.8	7.9	5.3	5.5	−6.6	−21.7	−5.8	−16.3
8 km/h	8.1	8.8	6.1	4.7	−10	−27	−7.8	−21.6
10 km/h	13	13	9.7	9	−13.7	−33.5	−11.4	−33.6

1.2.2 Accelerations at Other Locations on the Body

An additional set of acceleration measurements was performed, in order to compare the accelerations acting on the shoe during leg swing, with other body parts. As an example, the accelerations measured on the x-axis at the ankle are shown in Fig. 1.11a. In Fig. 1.11b from [37] the RMS acceleration measured at four motion speeds and five different body locations is shown. The locations include the ankle, the knee, the hip, the elbow and the wrist. At standstill with the arms hanging down, the measurement axis is oriented horizontally and points forward. The RMS value of the complete acceleration signal (not only the average of the peaks as was the case in Table 1.3) is calculated as an indicator of which part of the body performs more motion during walking.

The reader may notice that the lower body parts, specifically the ankle and knee show the greatest potential for energy harvesting, independent of the motion speed. The motion of the hip and elbow is balanced during walking and comparatively low. The motion of the wrist increases greatly as the runner transitions from walking to jogging and faster arm motion occurs.

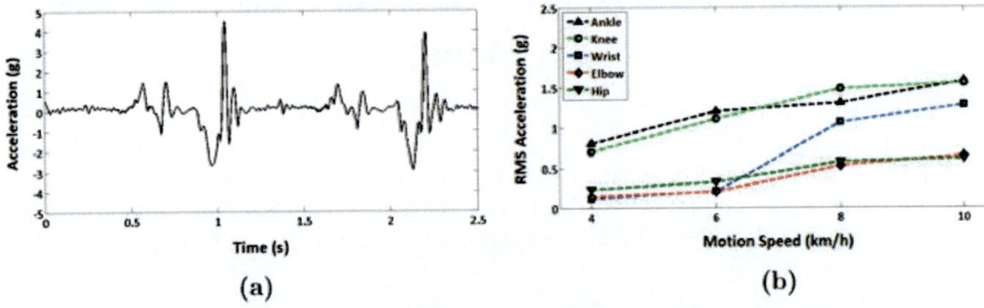

Fig. 1.11 **a** Accelerations recorded at the ankle on the x axis at a motion speed of 4 km/h for test person "1". **b** The RMS acceleration of several body parts, as calculated from the complete acceleration signal for test person "1". Figure 1.11a from [37]

Throughout this work, the acceleration data measured at the shoe is used as an input into the system simulations. All simulations are performed for the case of a motion speed of 4 km/h unless noted otherwise. From the available acceleration data set for test person “1” and the four motion speeds of 4, 6, 8 and 10 km/h, the case of 4 km/h is considered the “worst case” from the energy point of view. The slow walking speed of 4 km/h exhibits low acceleration peaks, which in turn translate into a low excitation of a harvesting device and a low power output. The experimental results are handled similarly. Measurement results, e.g. voltage and power output graphs are mostly given for the case of 4 km/h unless noted otherwise.

1.3 Basic Principle and Boundary Conditions

The previous section explains, how the accelerations from the swing motion of the foot during walking can be used as an energy source for energy harvesting. Inductive energy harvesting is based on the relative motion between magnets and coils due to inertia. The varying magnetic field due to magnet motion is coupled into the coils and induces a voltage therein. These components are the transducer of the system as electrical energy is generated out of kinetic energy. This transduction mechanism is the basis of this work and is explained in detail in Chap. 2.

In order to find device types, that can use the swing motion as an excitation source, the boundary conditions need to be defined first. The available space is the key constraint in this work. The magnet-coil arrangement has to be integrated into a shoe sole in a manner that allows the relative motion between magnets and coils. An x-ray picture of a running shoe with an integrated harvester is shown in Fig. 1.12. Clearly, the device size is limited by the sole of the shoe. The device height can under no circumstance be as large as the shoe sole itself, as material is needed below and above the harvesting device to protect it from environmental factors and the impact with the ground during walking as well as to protect it from the weight of the user. The sole material above and below the shoe is also important in terms of the wearing comfort as the spring-like characteristic of the sole is reduced significantly by the rigid harvesting device. With the device height being so restricted, it is not reasonable to harvest energy from the vertical accelerations that occur during walking. The height restriction would severely limit the possibility of relative motion between magnets and coils.

Because no relevant accelerations occur in a direction sideways from the body (along the z-axis in Fig. 1.9), the only feasible excitation direction is along the x-axis in Fig. 1.9 corresponding to a forward motion of the foot. The length of a harvesting device within a shoe sole is not as restricted as the height as can be seen in Fig. 1.12. The device cannot be as long the whole sole though. On the one hand, the shoe sole gets thinner towards the toes, allowing only the section under the heel as an integration point. On the other hand, the sole bends significantly in the front region of the shoe, so that a rigid device cannot be integrated there.

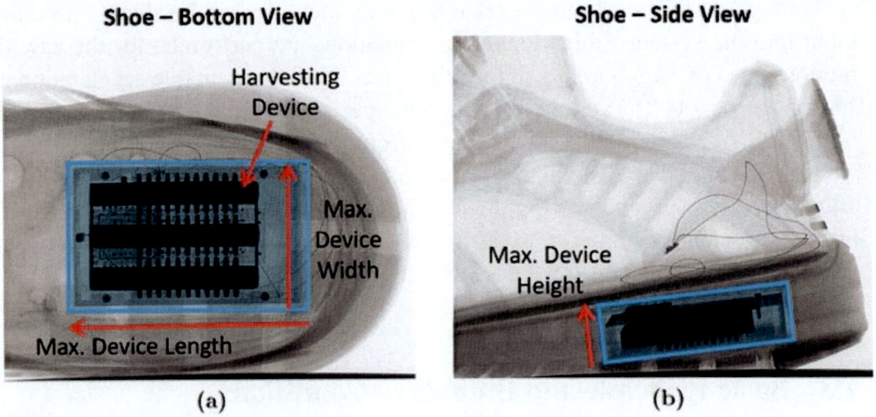

Fig. 1.12 X-ray pictures of a running shoe with an integrated energy harvesting device in side view (**a**, modified from [37].) and bottom view (**b**). The space that is available for the integration of a device into the shoe sole is marked

Table 1.4 Size constraints for the simulations and prototypes fabricated in this work

	Length	Width	Height
Housing	77 mm	41.5 mm	15.75 mm
Transduction mechanism	71 mm	37.5 mm	12.5 mm

In terms of width, the restriction is governed by the requirement of the structural stability of the shoe, i.e. the amount of sole material surrounding the shoe along its sides. For this work the size constraints listed in Table 1.4 were used. Originally, a length of 75 mm, width of 40 mm and height of 15 mm were chosen for the housing after discussions with a research institute specialized in shoe technologies and their durability analysis. The size of the transduction mechanism was chosen as listed in Table 1.4, which resulted in very thin walls of the housings. These proved to be not very durable and broke on several occasions during experimental testing. Consequently, the housing walls were strengthened to withstand the weight of the user, leading to the final device dimensions in Table 1.4.

Given the confined building space in the shoe sole, the choice of a transducer arrangement is limited. Two possible arrangements are given in Fig. 1.13. The first option is for the transducer to span the whole width of the device. Since the height is far smaller than the width, this cannot be achieved with a cylindrical magnet-coil-arrangement. A flat rectangular system is chosen to span the whole width of the building space.

If the available space is subdivided along the width, cylindrical setups, e.g. where the magnets can move within the coils, become another reasonable approach. the maximum transducer width of 37.5 mm equals three times the transducer height of 12.5 mm. Theoretically it is possible to fit three cylindrical transducers with a maxi-

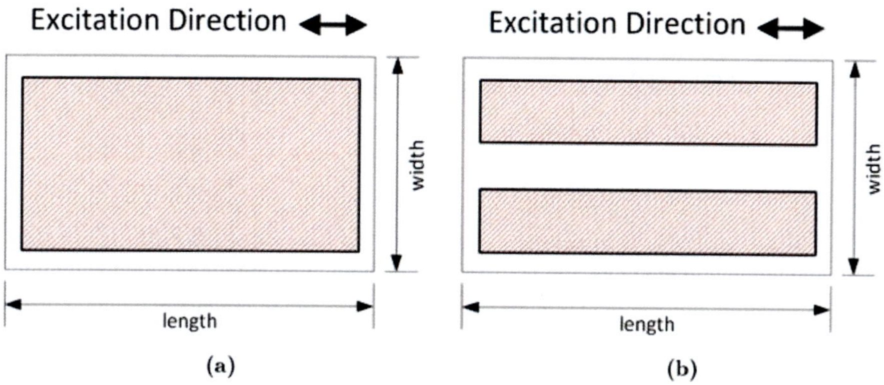

Fig. 1.13 Schematics of the available building space in the shoe sole. Possible transducer arrangements along the full width of the housing in (**a**) and as two separate transducers in (**b**) are marked in red. Figures from [34]

mum diameter of 12.5 mm within one housing. Due to the attractive forces between magnets however, it is expected, that a maximum of two cylindrical transducers can be placed in the housing as shown in Fig. 1.13b, without a negative influence on their respective motion. These descriptions are the basis for the four harvester architectures that are investigated in detail in this work. The following section gives a first overview of these four architectures.

1.4 EH Architectures

In this work a total of four harvester architectures (HAC), i.e. different magnet-coil-configurations are investigated. The HACs differ from one another by the shape of the magnets and coils and whether the magnets move within the coils or past the outside of them. The number of magnets and coils within a HAC, as well as the geometrical parameters of the components, are not fixed. The objective of the system model is to find the optimum configuration and parameters for each HAC. A schematic depiction of each architecture is given in Fig. 1.14 from [38]. The architectures in Fig. 1.14a, b are cylindrical systems. The dashed line in the figure indicates the axis of symmetry. These two architectures contain axially polarized magnets. In theory it is possible to use radially polarized magnets (giving HAC2 and HAC4). In practice however, it is not possible to fabricate, or more precisely to radially polarize magnets of such a small size [39]. Consequently, no experimental evaluation can be performed for HAC2 and HAC4 and the model cannot be verified for those particular cases. These two architectures therefore are not considered any further.

In HAC1 the magnets can either be full cylinders or rings with an axial polarization. Neighboring magnets are placed in repulsing orientation with a steel spacer of equal radius between them. The field of the opposing poles is guided through the

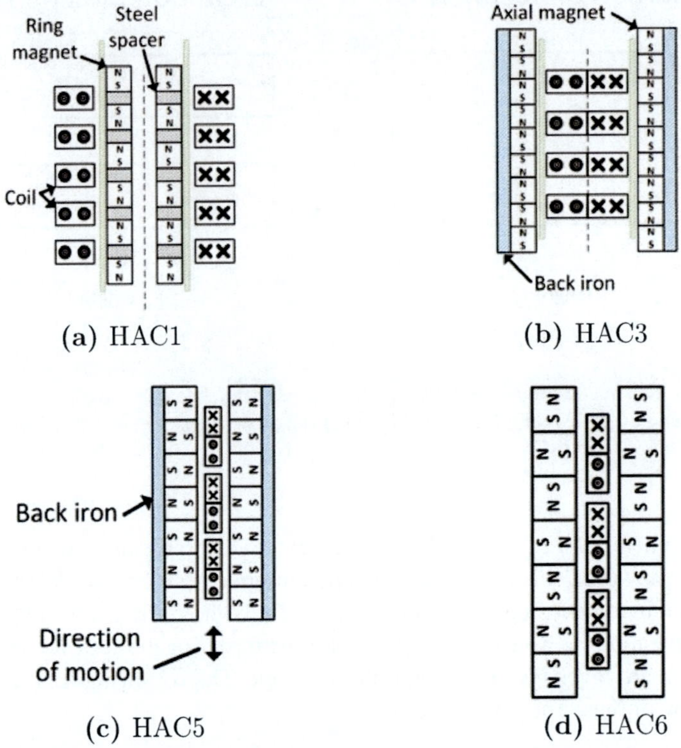

Fig. 1.14 Schematics of investigated swing harvesters HAC1 to HAC6. The dashed line indicates the axis of symmetry for the axisymmetric (cylindrical) architectures in (**a**) and (**b**). The rectangular architectures are shown in (**c**) and (**d**). All figures are shown in cross-sectional view. The direction of motion shown in (**c**) is equal across all architectures. Figure modified from [38]

spacer and towards the coils which are placed outside the plastic channel through which the magnets move (shown in light green). In HAC3 the coils are also fixed within the housing, but, contrary to HAC1, the axially polarized ring magnets move outside the coils. In this case only field components within the magnet ring contribute to the induced voltage in the coils. As was mentioned above, HAC2 and HAC4 are variations of HAC1 and HAC3 respectively with a radial polarization, which are not considered any further.

HAC5 and HAC6 are cuboid architectures which form a closed magnetic circuit around the coils. In both cases the coils are placed within a printed circuit board (PCB), which is fixed within the housing. The magnetic circuit encloses the PCB and can move freely within the housing. In HAC5 the magnets are placed in an attractive orientation on either side of the coils and a small steel plate is used to redirect the magnetic flux to the neighboring magnet. It is also possible to orient the magnet polarizations of HAC5 in such a way, that magnets within a side of HAC5 oppose each other (i.e. the magnetic flux direction runs along the length of HAC5,

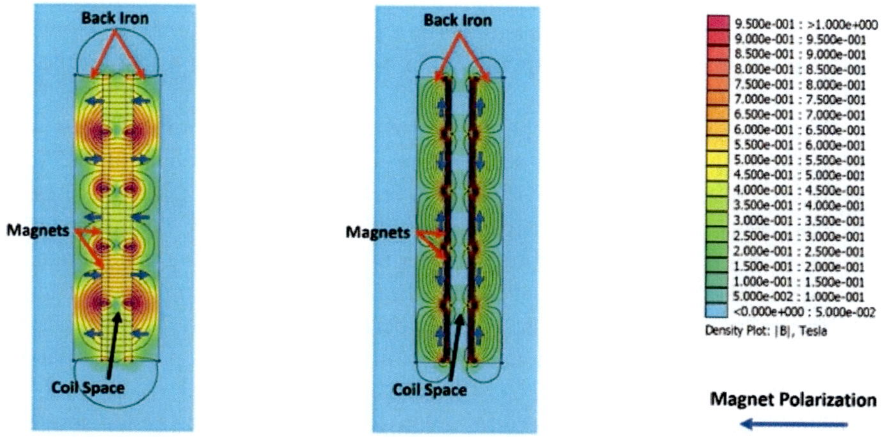

(a) HAC5 Original (b) HAC5 Variation (c) FEM Legend

Fig. 1.15 FEM simulation results for HAC5 in (**a**) and a variation of HAC5 with the magnet polarization shifted by 90° in (**b**). The legend is shown in (**c**). The magnetic flux density in the coil space is far larger for the original HAC5 compared to its variation. Figures from [34]

see Fig. 1.15b) and thus they do not face the opposing side of HAC5. This is a magnet orientation that resembles HAC1. The FEM simulations in Fig. 1.15b show that the flux density within the two magnet groups, i.e. in the coil space, is far smaller than for the standard HAC5 shown in Fig. 1.15a.

Consequently, this possible modification of HAC5 is not considered any further in this work. HAC6 forms a double "Halbach Array". The magnets are placed in a specific order, to generate a large magnetic flux density within the two parts of the Halbach array, i.e. in the coils. The number of magnets required to form such an array however is increased due to the use of magnets which solely serve to redirect the magnetic field and do not directly contribute to the field that penetrates the coils. Due to this particular arrangement, the flux density outside the Halbach array is greatly reduced, thus inherently shielding the device [40]. HAC5 is also shielded by means of the steel flux guides. The Halbach array can also be realized in a cylindrical arrangement, both as a single and as a double Halbach array [41]. However, as was the case with HAC2 and HAC4, radially polarized ring magnets cannot be produced at the small dimensions required in this work, thus this possible architecture is not considered herein.

1.5 Objectives of This Work and Major Achievements

The aim of this work is to lay the groundwork for the development of inductive energy harvesting devices that produce the maximum possible power output for the swing motion of the leg during walking as the given excitation source. The integration into the shoe sole poses very strict constraints on the device size, therefore the geometrical device parameters have to be optimized for a maximum power output for the given space.

The basis for a parameter optimization is the development of a flexible system model. Based on a real-world acceleration data input, the model can predict the power output of a given inductive harvester architecture. The magnetic flux distribution of the moving magnet arrangement needs to be calculated. The magnets serve as an inertial mass and their motion is determined by considering the external excitation as well as other significant effects like friction and electromagnetic damping. The calculation of the flux distribution is performed separately for each harvester architecture, but the system model itself is independent of the particular harvester architecture.

Four swing harvester architectures are investigated. Using the system model, the geometrical parameters for each device are optimized and once fabricated, the devices are evaluated experimentally during treadmill experiments. The experimental data is used to revise the system model and a good agreement between simulations and experiments is achieved. A second generation of devices is fabricated, which achieve a far greater output than the first generation. Experiments in industrial applications reveal, that these harvester types can be used anywhere, where the excitation is non-resonant and where the amplitudes are large. The developed model can now be used to investigate further architectures, or to optimize the power output of a particular harvester setup depending on the application, i.e. the size constraints and the given excitation.

1.6 Organization of the Book

Before actual harvesting devices of a specific architecture can be fabricated, the geometrical parameters, e.g. arrangement, size and number of magnets and coils, need to be determined. In order to find suitable sets of geometrical parameters, first a theoretical system model is developed as shown in Chap. 2. This model is based on a differential equation of motion and considers the forces that act on the moving magnet stacks. The mechanical motion model was designed to be independent of the harvester architecture, while the calculation of the magnetic field and the coupling coefficient are architecture-dependent. The system model is used to predict the power output of the devices. It enables the optimization of devices for a given excitation and building space, thus using the given volume very efficiently.

Once the system model is available, the geometrical parameters for each architecture are optimized for a maximum power output as described in Chap. 3. Harvesting devices are then fabricated according to the calculated parameters. Subsequently they are characterized experimentally and the measurement results are used to verify and revise the system. This is presented in Chap. 4. The predicted power outputs prior to the model revision deviated significantly from the experimental data. With the revised model, the power output predictions are expected to be more reliable in future simulations.

Building on the revised model, a new set of optimizations is performed in Chap. 5 in order to determine the best set of geometrical parameters for each of the four investigated architectures. These new parameters set the basis for the second generation of harvesting devices which show a greatly increased power output as explained in Chap. 6.

In Chap. 7 experiments and results for two industrial applications are shown. A pneumatic piston is investigated as a source of linear excitation that can be reproduced in a laboratory environment. The harvesting devices are characterized on the piston. In a second application, measurement results from a harvester placed on an industrial clutch-brake are shown. Additionally, a human-body application is described briefly. This is the concluding block of this work, which proves that the devices developed in this work can be used beyond human-body applications, in industrial environments where linear accelerations are present.

Lastly, the conclusion is given in Chap. 8, along with an outlook on future work building on this work.

References

1. D. Hoffmann, A. Willmann, B. Folkmer, Y. Manoli, Tunable vibration energy harvester for condition monitoring of maritime gearboxes, in *PowerMEMS* (IOP, ed.), vol. 557, p. 012099 (2014)
2. M.-K. Kim, M.-S. Kim, S. Lee, C. Kim, Y.-J. Kim, Wearable thermoelectric generator for harvesting human body heat energy. Smart Mater. Struct. **23**(10), 105002 (2014)
3. P. Niu, P. Chapman, R. Riemer, X. Zhang, Evaluation of motions and actuation methods for biomechanical energy harvesting, in *Power Electronics Specialists Conference, 2004. PESC 04. 2004 IEEE 35th Annual*, vol. 3, pp. 2100–2106 (2004)
4. A. Cervera, Z. Rubinshtein, M. Gad, R. Riemer, M.M. Peretz, Biomechanical energy harvesting system with optimal cost-of-harvesting tracking algorithm. IEEE J. Emerg. Sel. Top. Power Electron. **4**(1), 293–302 (2016)
5. L. Xie, R. Du, Harvest human kinetic energy to power portable electronics. J. Mech. Sci. Technol. **26**(7), 2005–2008 (2012)
6. T. Xue, X. Ma, C. Rahn, S. Roundy, Analysis of upper bound power output for a wrist-worn rotational energy harvester from real-world measured inputs. J. Phys.: Conf. Ser. **557**, 012090 (2014)
7. Q. Li, M. Shepertycky, J.-P. Martin, Lower-limb driven energy harvesting backpack: design, performance and energetics, in *2015 34th Chinese Control Conference (CCC)*, pp. 7821–7826 (2015)
8. Z. Yang, A. Khaligh, A flat linear generator with axial magnetized permanent magnets with reduced accelerative force for backpack energy harvesting, in *2012 Twenty-Seventh Annual IEEE Applied Power Electronics Conference and Exposition (APEC)*, pp. 2534–2541 (2012)
9. J. Granstrom, J. Feenstra, H.A. Sodano, K. Farinholt, Energy harvesting from a backpack instrumented with piezoelectric shoulder straps. Smart Mater. Struct. **16**(5), 1810–1820 (2007)
10. J. Feenstra, J. Granstrom, H. Sodano, Energy harvesting through a backpack employing a mechanically amplified piezoelectric stack. Mech. Syst. Signal Process. **22**(3), 721–734 (2008)

11. Y. Cha, S. Hong, Energy harvesting from walking motion of a humanoid robot using a piezoelectric composite. Smart Mater. Struct. **25**(10), 10LT01 (2016)
12. M. Pozzi, Magnetic plucking of piezoelectric bimorphs for a wearable energy harvester. Smart Mater. Struct. **25**(4), 045008 (2016)
13. W. Wang, J. Cao, N. Zhang, J. Lin, W.-H. Liao, Magnetic-spring based energy harvesting from human motions: design, modeling and experiments. Energy Convers. Manag. **132**, 189–197 (2017)
14. K. Ylli, D. Hoffmann, A. Willmann, B. Folkmer, Y. Manoli, Investigation of pendulum structures for rotational energy harvesting from human motion. J. Phys.: Conf. Ser. **660**, 012053 (2015)
15. J. Kymissis, C. Kendall, J. Paradiso, N. Gershenfeld, Parasitic power harvesting in shoes, in *Digest of Papers. Second International Symposium on Wearable Computers*, 19–20 Oct., pp. 132–139 (1998)
16. K. Ishida, T.-C. Huang, K. Honda, Y. Shinozuka, H. Fuketa, T. Yokota, U. Zschieschang, H. Klauk, G. Tortissier, T. Sekitani, H. Toshiyoshi, M. Takamiya, T. Someya, T. Sakurai, Insole pedometer with piezoelectric energy harvester and 2 v organic circuits. IEEE J. Solid-State Circuits **48**(1), 255–264 (2013)
17. A. Gatto, E. Frontoni, Energy harvesting system for smart shoes, in *2014 IEEE/ASME 10th International Conference on Mechatronic and Embedded Systems and Applications (MESA)*, pp. 1–6, IEEE, Senigallia (2014)
18. W.G. Li, S. He, S. Yu, Improving power density of a cantilever piezoelectric power harvester through a curved l-shaped proof mass. IEEE Trans. Ind. Electron. **57**(3), 868–876 (2010)
19. W. Zeng, X.-M. Tao, S. Chen, S. Shang, Chan, Helen Lai Wah, S. H. Choy, Highly durable all-fiber nanogenerator for mechanical energy harvesting. Energy Environ. Sci. **6**(9), 2631 (2013)
20. R. Kornbluh, R. Pelrine, Q. B. Pei, R. Heydt, S. Stanford, S. J. Oh, J. Eckerle, Electroelastomers: Applications of dielectric elastomer transducers for actuation, generation and smart structures, *Smart Structures and Materials 2002: Industrial and Commercial Applications of Smart Structures Technologies*, vol. 4698, pp. 254–270 (2002)
21. C. Jean-Mistral, S. Basrour, J.-J. Chaillout, Modelling of dielectric polymers for energy scavenging applications. Smart Mater. Struct. **19**(10), 105006 (2010)
22. T. Krupenkin, J.A. Taylor, Reverse electrowetting as a new approach to high-power energy harvesting. Nature Commun. **2**, 448 (2011)
23. T.-H. Hsu, S. Manakasettharn, J.A. Taylor, T. Krupenkin, Bubbler: a novel ultra-high power density energy harvesting method based on reverse electrowetting. Sci. Rep. **5**, 16537 (2015)
24. P. Bai, G. Zhu, Z.-H. Lin, Q. Jing, J. Chen, G. Zhang, J. Ma, Z.L. Wang, Integrated multilayered triboelectric nanogenerator for harvesting biomechanical energy from human motions. ACS Nano **7**(4), 3713–3719 (2013)
25. H. Li, L. Su, S. Kuang, Y. Fan, Y. Wu, Z.L. Wang, G. Zhu, Multilayered flexible nanocomposite for hybrid nanogenerator enabled by conjunction of piezoelectricity and triboelectricity. Nano Res. **10**(3), 785–793 (2017)
26. D. Hoffmann, B. Folkmer, Y. Manoli, Human motion energy harvester for biometric data monitoring. J. Phys.: Conf. Ser. **476**, 012103 (2013)
27. L. Moro, D. Benasciutti, Harvested power and sensitivity analysis of vibrating shoe-mounted piezoelectric cantilevers. Smart Mater. Struct. **19**(11), 115011 (2010)
28. M. Salauddin, J.Y. Park, Design and experiment of human hand motion driven electromagnetic energy harvester using dual halbach magnet array. Smart Mater. Struct. **26**(3), 035011 (2017)
29. Y. Rao, K.M. McEachern, D.P. Arnold, A compact human-powered energy harvesting system. J. Phys.: Conf. Ser. **476**, 012011 (2013)
30. M.A. Halim, J.Y. Park, A miniaturized human-motion energy harvester using flux-guided magnet stacks. J. Phys.: Conf. Ser. **773**, 012089 (2016)
31. N. Shenck, J. Paradiso, Energy scavenging with shoe-mounted piezoelectrics. IEEE Micro **21**(3), 30–42 (2001)

32. J.-X. Shen, C.-F. Wang, P.C.-K. Luk, D.-M. Miao, D. Shi, C. Xu, A shoe-equipped linear generator for energy harvesting. IEEE Trans. Ind. Appl. **49**(2), 990–996 (2013)
33. D. Carroll, M. Duffy, Modelling, design, and testing of an electromagnetic power generator optimized for integration into shoes. Proc. Inst. Mech. Eng. Part I: J. Syst. Control. Eng. **226**(2), 256–270 (2012)
34. K. Ylli, Energy harvesting from the swing motion of the human leg (2019)
35. N.A. Borghese, L. Bianchi, F. Lacquaniti, Kinematic determinants of human locomotion. J. Physiol. -Lond. **494**(3), 863–879 (1996)
36. K. Ylli, D. Hoffmann, A. Willmann, P. Becker, B. Folkmer, Y. Manoli, Energy harvesting from human motion: exploiting swing and shock excitations. Smart Mater. Struct. **24**(2), 025029 (2015)
37. K. Ylli, D. Hoffmann, P. Becker, A. Willmann, B. Folkmer, Y. Manoli, Human motion energy harvesting for aal applications. J. Phys.: Conf. Ser. **557**, 012024 (2014)
38. K. Ylli, D. Hoffmann, A. Willmann, B. Folkmer, Y. Manoli, Human motion energy harvesting: numerical analysis of electromagnetic swing-excited structures. Smart Mater. Struct. **25**(9), 095014 (2016)
39. K. Ylli, Fertigung radial polarisierter magneten: e-mail (2015)
40. D. Zhu, S. Beeby, J. Tudor, N. Harris, Vibration energy harvesting using the halbach array. Smart Mater. Struct. **21**(7), 075020 (2012)
41. I. Shahosseini, K. Najafi, Cylindrical halbach magnet array for electromagnetic vibration energy harvesters, in *2015 28th IEEE International Conference on Micro Electro Mechanical Systems (MEMS)*, pp. 1051–1054

Chapter 2
Theory and Modeling

With today's technical progress in simulation, the physical behavior of a system is analyzed using modern computers and software, before its actual fabrication. A car chassis for example is crashed digitally to find weak spots in advance. The same method is applied to the development of energy harvesting (EH) devices. EH devices are highly application-specific, e.g. due to the characteristic frequencies of the vibrations of a certain machine. In order to maximize the power output of an EH system, the optimal parameter combination of components for a given system volume has to be determined. In the case of inductive harvesters this includes the size and number of the used magnets and coils amongst others.

A system model is developed to represent the behavior of the EH by considering significant physical effects which influence the system, or to be more precise, which influence the motion of the moving magnets relative to the coils. For a given excitation, for example acceleration data from the human gait, and a given set of geometrical parameters, the system model predicts the motion of the magnets and consequently the power output of the device. This chapter describes the development of a system model for non-resonant inductive harvesters which uses acceleration data from the foot as an excitation source. The model is implemented in the Matlab/Simulink environment.

The system model is based on a differential equation of motion (eom), which describes the relation between the forces in the system and the magnet motion. The eom is described in detail in Sect. 2.1. Details on the simplified calculation of the friction force are given in Sect. 2.2. The mechanical part of the system model also considers the hardstops of the system. The implementation of the hardstops is given in Sect. 2.3. The electrical damping force (edf) is the force that results when a magnet induces a voltage in a coil. The corresponding current flow in the coil windings generates a magnetic field that opposes its cause, i.e. the magnet motion. In order to determine the edf, the magnetic flux density of the arrangement of permanent magnets is calculated as shown in Sect. 2.4. The electromagnetic coupling function is determined as described in Sect. 2.5 and gives the relation between the magnet

K. Ylli and Y. Manoli, *Energy Harvesting for Wearable Sensor Systems*,
Springer Series in Advanced Microelectronics 62,
https://doi.org/10.1007/978-981-33-4448-8_2

speed and the induced voltage for a given magnet-coil arrangement. Based on these calculations the induced voltage, edf and power output of the system are calculated as shown in Sect. 2.6.

2.1 Differential Equation of Motion

The type of inductive harvester that is developed in this work is based on a magnetic proof mass. This proof mass is a stack of several magnets that can slide freely within a housing and induce a voltage in coils that are fixed within the housing according to Faraday's law of induction. The harvesters are designed as flat devices according to the restrictions described in Sect. 1.3, therefore the motion of the magnets that form the proof mass m is caused solely by the external acceleration $a_{h,ext}$ of the foot as indicated in Fig. 2.1. The modeling done in this work can be applied to all investigated architectures without loss of generality. In order to determine the induced voltage within the coils, the relative motion between the magnet stack and the coils has to be calculated.

The analogy of a tablecloth that is swiftly pulled from the table without the glasses falling off can be used to explain the magnet motion. As long as the external acceleration applied by the foot to the housing $a_{h,ext}$ is smaller than a certain threshold acceleration a_{thr}, the proof mass m will move at the same global velocity as the housing due to friction between the two bodies. Accelerations are used analogously to forces in this argument, because the input to the system model is provided as acceleration data.

In fact, there is a balance of forces between the inertial force $F_T = m \cdot a$ and the static friction force. The expression $a_{h,ext} < a_{thr}$ is thus equivalent to $F_T = F_{brk}$, i.e. no motion of the magnet in the local coordinate system (LCS) of the housing (F_{brk} is the breakaway friction force which is described in more detail in Sect. 2.2).

Consequently the relative motion between the two bodies (moving magnets and fixed coils in the housing) is zero and no voltage is induced in the coils. When the

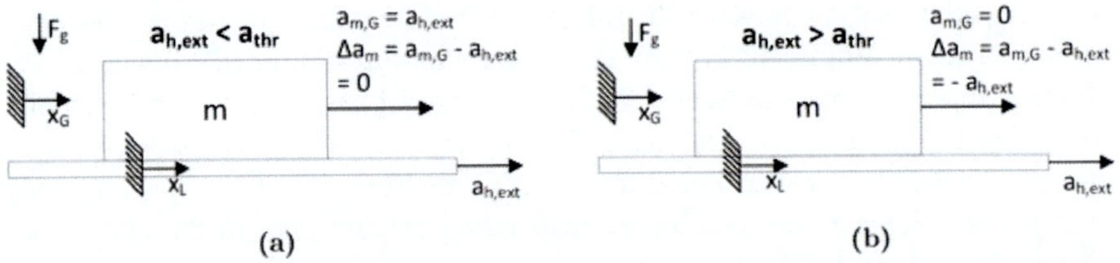

Fig. 2.1 Proof mass m resting on a plate (e.g. housing) that experiences the external acceleration $a_{h,ext}$. **a** Due to the stick friction force, at accelerations smaller than the threshold a_{thr} the proof mass moves with the plate at zero relative acceleration between the two bodies. **b** At large accelerations, the plate pulls away under the proof mass at the relative acceleration $\Delta a_m = -a_{h,ext}$ in the ideal case without kinetic friction. Figures from [9]

inertial force is larger than the breakaway friction force ($F_T > F_{brk}$), the external acceleration $a_{h,ext}$ is larger than a_{thr} and the magnets will tend to remain in their position within the global coordinate system (GCS) x_G, while the housing pulls away underneath it with $a_{h,ext}$ as the stick friction force is overcome.

The magnet stack experiences a global acceleration $a_{m,G}$ of 0 g in the GCS in the ideal case without friction (in the system model $a_{m,G}$ depends on the kinetic friction force). A relative motion between the bodies is thus created and the changing magnetic flux density induces a voltage in the coils. From these explanations, it follows that the magnets perform a relative backward motion within the LCS of the housing as the foot moves forward, while analogously, the magnets move forward within the housing as the foot decelerates before the stance phase.

The motion of the magnets relative to the housing and thus the coils, leads to the induction of a voltage. In order to be able to predict the power output of any particular HAC, this motion has to be determined first. This can be achieved using the differential equation of motion given in Eq. 2.1.

$$-m_{mag}\ddot{x} = m_{mag}A_{ext} + F_R(\dot{x}) + F_{el}(\dot{x}, x) + F_S(x) \tag{2.1}$$

The motion of the magnet stack is described by $x, \dot{x}, \ddot{x}$ in the LCS where m_{mag} is the mass of the magnet stack (including steel flux guides, where applicable). The acceleration $\ddot{x}$ of the moving magnets follows from the sum of the forces that influence the motion. The measured acceleration of the foot A_{ext} is the only "external" influence on the magnet motion.

Other effects include the friction force F_R (which contains both static and kinetic components), the electrical damping force F_{el} and the spring force F_S which provides a restoring force, when springs are used to stop the magnets at the end of their motion range. The reader may notice from Eq. 2.1 that the external acceleration does not depend on the magnet motion. The friction force F_R only depends on the motion speed of the magnet, while F_{el} depends on both the motion speed and the position of the magnet relative to the coils.

The block diagram in Fig. 2.2 shows all forces that contribute to the magnet motion as well as the external acceleration. The diagram is a simplified representation of the system model and shows the feedback paths of the system. The forces are dependent on the magnet motion and position and there is a force feedback. The forces are divided by the weight of the moving magnet to give an equivalent acceleration. The acceleration is then integrated to determine the speed of the magnet motion and the magnet position within the housing. The force due to springs, e.g. instead of hardstops, and due to air compression as the magnets push air within a tightly confined space during their motion, are not considered in this work.

The equation of motion is rearranged as shown in Eq. 2.2.

$$-\left(A_{ext} + \frac{F_R(\dot{x}) + F_{el}(\dot{x}, x) + F_S(x)}{m_{mag}}\right) = \ddot{x} \tag{2.2}$$

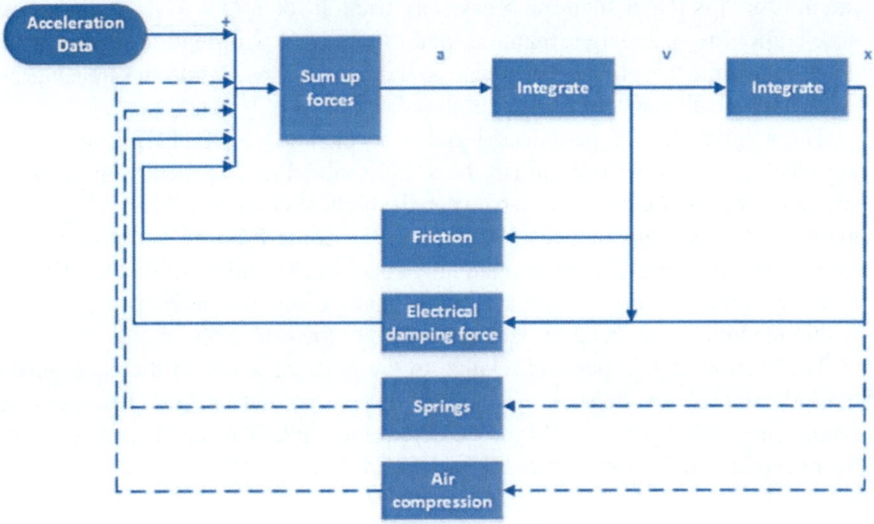

Fig. 2.2 General block diagram of the system model. The optional blocks that are not implemented in this work, i.e. restoring forces from springs and from the air compression due to magnets moving in an enclosed space, are connected by dashed lines. Figure from [9]

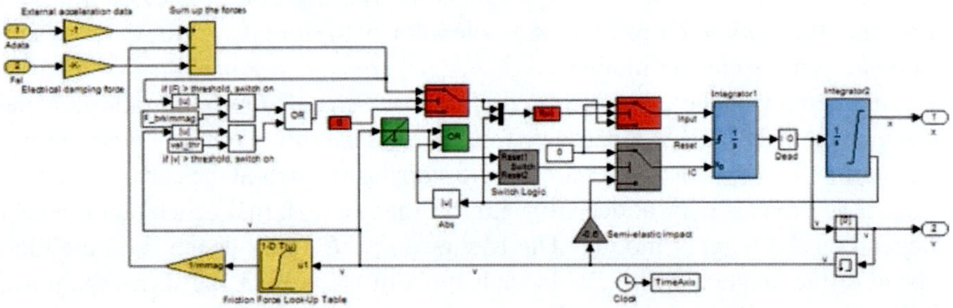

Fig. 2.3 The mechanical components of the system model are shown in the Simulink environment. Blue blocks are integrators, yellow blocks are the forces from the 'eom', green blocks trigger the integrator reset, red blocks set the integrator input $\ddot{x}$ and gray blocks set the output start value of INT1 after a reset. The magnet position and speed serve as inputs to the electromagnetic components that are described starting in Sect. 2.4. Figure from [10]

This is the basis of the numerical simulation model. By integration of the acceleration $\ddot{x}$, the speed and position of the magnet within the housing can be determined. The Simulink block diagram containing the mechanical components is shown in Fig. 2.3. The blocks have been highlighted according to their function.

The blue blocks are the integrators, which generate x and $\dot{x}$ from $\ddot{x}$. A saturation condition may be defined in the integrator, which in this case models the limited motion range of the magnet stack. A reset condition can be implemented to model the reversal of the motion direction upon reaching the channel end. The green blocks trigger the integrator reset, the red blocks set the integrator input $\ddot{x}$ and the gray blocks set the output start value of "Integrator1" (INT1) after a reset. The yellow blocks sum up the forces from the eom. The external acceleration data and the electrical damping force are generated in other parts of the model, while the friction force is implemented as a look-up table at the bottom left of Fig. 2.3.

The two main effects which influence the motion of the magnet stack are the friction between the magnet stack and housing as well as the hardstops, as the magnet reaches the end of its motion range. These key aspects of the system model are described in more detail in the following sections.

2.2 Friction Model

In this work a kinetic friction model based on Coulomb friction between two rigid bodies is implemented [1–3]. Three different states of friction are distinguished depending on the motion of the moving part as shown in Fig. 2.4.

The "rest" state applies to velocities smaller than a certain threshold velocity v_{thr}. In this region from $-v_{thr}$ to v_{thr} the friction force F_R is linear up to the breakaway friction force F_{brk} which marks the point where motion begins.

$$F_R(v) = F_{brk} \cdot \frac{v}{v_{thr}} \tag{2.3}$$

This linear region according to Eq. 2.3 is necessary due to the nature of numerical simulations which cannot properly react to abrupt increases or discontinuities in the calculated values [1, 2]. The breakaway friction is calculated according to Eq. 2.4.

$$F_{brk} = f_c \cdot v_{thr} + F_C + \left(F_{brk,S} - F_C\right) \cdot e^{-c_v \cdot v_{thr}} \tag{2.4}$$

where f_c is the viscous friction coefficient in N/(m/s), $F_c = m_{mag} \cdot g \cdot \mu_k$ is the Coulomb friction force, $F_{brk,S} = m_{mag} \cdot g \cdot \mu_s$ is the static breakaway friction force and c_v is the transition approximation coefficient in s/m.

The motion region ($v > v_{thr}$) may consist of up to three components. The Coulomb friction is constant for all velocities and is often used all by itself in simple friction models using the kinetic and static coefficients of friction μ_k and μ_s respectively. Throughout this work $\mu_k = 0.2$, $\mu_s = 0.3$, $f_c = 0.01$ N/(m/s), $v_{thr} = 0.001$ m/s and $c_v = 3$ s/m are used as a base set of parameters, while Sect. 2.7 deals with the uncertainties associated with some of the parameters used in the system model. The chosen coefficients of friction lie within the common range of values for friction between plastic and metal objects.

The threshold velocity is less than one percent of the magnet velocity that actually occurs in a fabricated device. The experimental magnet velocity is estimated from the time between voltage peaks measured for a given multi-magnet-arrangement that moves past a coil. The coefficients f_c and c_v are chosen based on the descriptions in [2]. For the devices developed in this work, the friction force is generally small (in

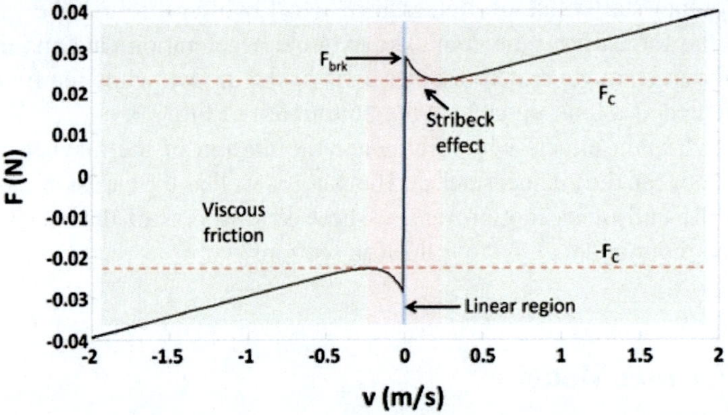

Fig. 2.4 Friction force curve calculated with the exemplary parameters $\mu_k = 0.2$, $\mu_s = 0.3$, $f_c = 0.01$ N/(m/s), $v_{thr} = 0.01$ m/s, $c_v = 10$ s/m and $m_{mag} = 10\,g$. This values were chosen to generate a curve which clearly shows the highlighted characteristic regions. Figure from [10]

the range of a few Newtons) due to the small magnet masses and motion speeds. As is shown in Sect. 4.5, a good match between the experimental and simulated power output is achieved with this friction setup.

The Stribeck friction marks the sloped decrease in friction force from the velocity threshold down to the Coulomb friction [1]. As the velocity of the moving part increases, the viscous friction $F_V = f_c \cdot v$ based on fluidic effects will dominate over the other components and cause an increase in the friction force. However, at the velocities present in the actual magnet motion, it is hardly noticeable. The friction force in the motion region is described by Eq. 2.5.

$$F_R(v) = \begin{cases} f_c \cdot v + F_C + \left(F_{brk,S} - F_C\right) \cdot e^{-c_v \cdot v} & v > v_{thr} \\ f_c \cdot v - F_C - \left(F_{brk,S} - F_C\right) \cdot e^{-c_v \cdot |v|} & v < v_{thr} \end{cases} \tag{2.5}$$

The friction force F_R is calculated within a range of reasonable velocities, e.g. $-5\,\text{m/s} \leq v \leq 5\,\text{m/s}$ and is implemented within the system model as a look-up table of the motion speed, where the friction force can be read as required. For each magnet geometry and thus for each magnet mass m_{mag} a new look-up-table needs to be calculated.

2.3 Mechanical Hardstops and Integrator Reset Conditions

The integrator blocks shown in Fig. 2.3 have to be adapted for the application at hand. As the motion of the magnet stacks is limited by the housing dimensions, the saturation condition of INT2 which generates x out of $\dot{x}$ has to be set. The maximum motion range within the housing is defined as x_{max}.

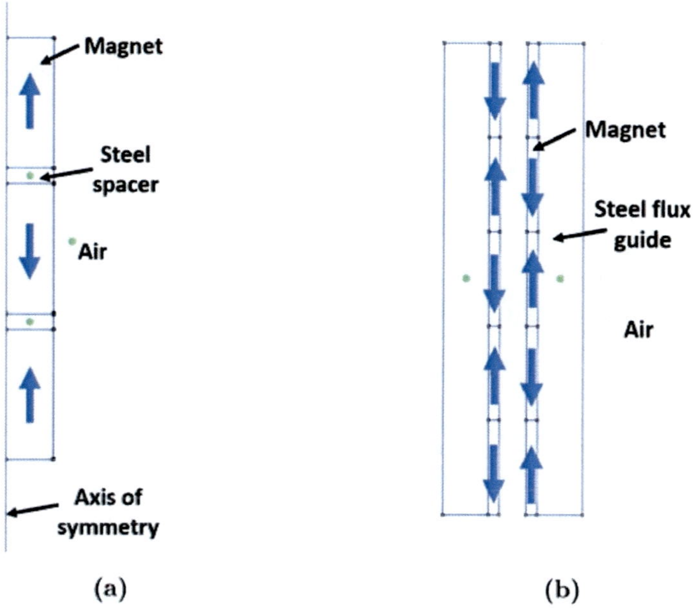

Fig. 2.5 Exemplary drawing of parameterized harvester structures within the FEM tool. **a** Shows the right half-plane of a cylindrical magnet stack with the axis of symmetry being represented by the left margin. **b** Shows the cross-sectional drawing of a rectangular magnet stack where no symmetrical relations can be exploited. The width of the magnet is directed into the drawing plane. Figures from [9]

The saturation of INT2 serves as a mechanical hardstop, which represents the turning point for the magnet stack, as it reaches the end of its travel range, e.g. the end of a cylindrical channel, during motion. INT2 provides a saturation signal at a dedicated port as the magnet reaches the travel boundary. Using this signal, a reset can be triggered (green blocks in Fig. 2.3). The input and output values of the integrator blocks can be overwritten upon such a reset, which allows a change in the direction of motion.

The position of the magnet stack is continuous, thus INT2 does not require a reset and only INT1 receives the saturation signal at its reset port. At that point, if a hardstop impact has occurred, a semi-elastic impact according to [4] is modeled with a coefficient of restitution of $k_i = 0.6$, i.e. the velocity output of INT1 after the reset will be 0.6 times the last output in the opposing direction (gray blocks in Fig. 2.3). However, if the magnet stack has reached the hardstop and is being pushed onto it by the acting forces, switches set the integrator input to zero until the sign of the forces changes, which indicates that motion in the opposing direction will take place (red blocks in Fig. 2.3). The logic Eq. 2.6 from the Simulink implementation (small red block $f(u)$ between the red switches in Fig. 2.3) represents this condition.

$$((u_2 = 1)\ \&\&\ (u_1 \geq 0))\ \|\ ((u_2 = -1)\ \&\&\ (u_1 \leq 0)) \tag{2.6}$$

The signal u_1 is the sum of all forces acting in the system and u_2 is the saturation signal from INT2. If the equation results in "True", the acting forces push the magnet onto one of the two hardstops and the speed of INT1, i.e. the integrator input port, is set to zero by the switch. Otherwise, if the forces no longer push the magnet onto the hardstop, the condition turns to "False" and the sum of the acting forces is routed to the input port of INT1.

2.4 Calculating the Magnetic Flux Using Finite Element Analysis

The calculation of the magnetic field distribution is hardly possible using an analytical approach as closed-form solutions exist only for a few limited magnet configurations. To the best of the author's knowledge, when ferromagnetic material distorts the magnetic field, closed-form solutions do not exist at all. Therefore a numeric approach using finite element analysis is used (Fig. 2.5).

All magnetic structures considered in this work are simulated using the free 2D FEM (finite element method) tool "FEMM" [5]. Two key advantages of this tool shall be named: firstly, it can be fully integrated into Matlab and run from within the Matlab environment. Secondly, every geometry under investigation can be parameterized within Matlab (or the similar freeware "Octave" for that matter), thus rendering the simulations highly flexible. In the case of the cylindrical structures, due to rotational symmetry, the magnetic setup is drawn on a half-plane as shown in Fig. 2.5a. The left margin of the image represent the rotational axis of symmetry. The rectangular architectures are drawn in a cross-sectional plane as shown in Fig. 2.5b. The magnetic field is assumed to be homogeneous along the depth of the rectangular magnets. While this is a simplification of the real setup and fringe effects are neglected, it enables the parameterized simulation at comparatively low computational cost.

In all simulations neodymium iron boron (NdFeB) permanent magnets with the polarization N35 (corresponds to a coercive field strength of $H_c = 868\,{}^{kA}/_{m}$ and assuming a simplified relative permeability of $\mu_{mag} = 1$) are used. While there are magnets with stronger polarization, the N35-type magnets are the most widely available.

The magnetic flux density is only extracted for the component perpendicular to the coil cross-section, which contributes to the induced voltage, i.e. B_z in the cylindrical case and B_x in the rectangular case. A matrix of flux density values B is constructed for all coordinates within the motion range of a magnet stack. This allows the calculation of the induced voltage for all positions of the magnets relative to the coils. The spacing between points in the B-matrix corresponds to the diameter of the copper wire used to wind the coils. If not mentioned otherwise, a wire diameter of $d_{Cu} = 0.1$ mm is used in this work.

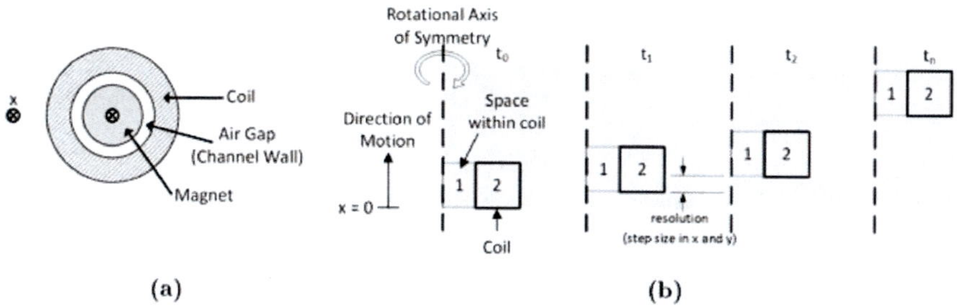

Fig. 2.6 **a** Schematic top view of an axis-symmetric magnet and coil setup. **b** Schematic of the calculation of the magnetic flux within a coils' winding for every position the coil can take relative to the magnets used in the system. The dotted gray area represents the magnetic flux distribution as a two-dimensional data matrix. Figures from [10]

The motion of the magnets relative to the coils, i.e. the change in magnetic flux density experienced by the coils, leads to induction of a voltage within the coils. Apart from the speed of the magnet and the position within the housing at any given time during the simulation, the flux density within the coil needs to be calculated in order to determine the electromagnetic coupling function $k(x)$. The coupling function $k(x)$ is calculated from the gradient of the magnetic flux φ and the position x according to Eq. 2.7.

$$k(x) = \frac{d\varphi}{dx} \tag{2.7}$$

Figure 2.6 shows a schematic cross-sectional view of an exemplary axis-symmetric architecture and the corresponding method to determine the flux density within the coils. It was mentioned before, that the flux density values are stored in a 2D matrix, as represented by the gray area in the figure. The coils can "move" from the bottom of this matrix to the top, covering all possible relative positions between coil and magnet. In order to generate a look-up table of coupling coefficients, the average flux density experienced by the coil at any given position along the direction of motion is determined.

This operation is not performed for the full coil as depicted in Fig. 2.6, but for each winding n in the radial direction of the coil (see Fig. 2.7). Due to the symmetry of axisymmetric architectures, only the right half-plane is considered. Boxes are spanned which represent the cross-sectional area covered by the coil windings of radius r_i. All matrix data points within the scope of this box (e.g. the highlighted blue box for the smallest windings of the coil in Fig. 2.7) are used to give the mean flux density B_{mean} for windings of that particular radius and position along the direction of motion.

In the next step, B_{mean} is calculated for the next windings of radius $r_{i+1} = r_i + d_{Cu}$ (represented by the sum of the highlighted blue and green boxes in Fig. 2.7) for $i = 1..n$ where n is the number of windings in the radial direction.

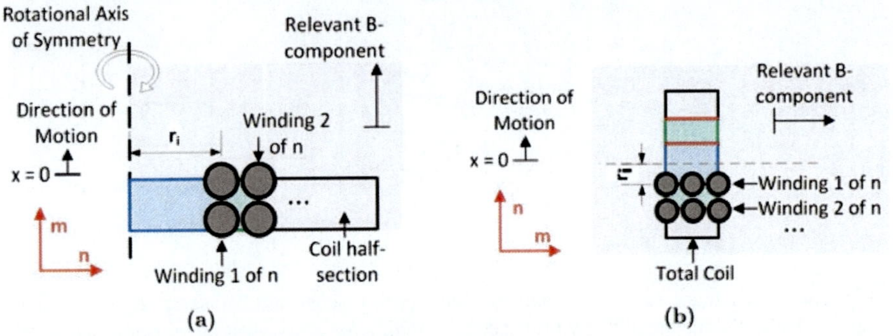

Fig. 2.7 Schematic showing the calculation of the mean flux density within the radial windings of a coil for an exemplary cylindrical architecture in (**a**) and a rectangular architecture in (**b**). The gray area represents the magnetic flux distribution as a two-dimensional data matrix. Single windings are indicated as black circles. Figures from [9]

In the case of rectangular HACs, rectangular coils are used to maximize the use of the air gap in between the two sides of the magnetic circuit. In practice, the coils are not exactly rectangular and show a slightly oval shape, but in the model they are implemented using rectangular windings. Again, owing to the 2D-simulation environment, the magnetic field is assumed to be homogeneous along the depth of the coil (into the drawing plane of Fig. 2.7b). While the direction of motion, i.e. the long edge of the housing, remains the same, the relevant component of the magnetic flux density is now perpendicular to the direction of motion due to the arrangement of the magnets (see Fig. 1.14). The method of the calculation of B_{mean} however remains the same.

2.5 Electromagnetic Coupling Function

The electromagnetic coupling creates the link between the induced voltage U_{ind} and the motion speed of the magnets as shown in Eq. 2.8 which is a rearrangement of Faraday's law of induction.

$$U_{ind} = -N \cdot \frac{d\varphi}{dt} = -N \cdot \left(\frac{dA}{dt} B + \frac{dB}{dt} A\right)_{|A=const} = -N \cdot \frac{dB}{dt} A \cdot \frac{dx}{dx} = -N \cdot \frac{dB}{dx} A \cdot \frac{dx}{dt} = k \cdot \dot{x} \tag{2.8}$$

The change in magnetic flux $d\varphi$ stems from either the change of magnetic flux density dB in a coil winding or from the change of the effective winding area in a static field, e.g. because the winding is rotating. The area of the windings is assumed to be constant as the position of the coils is fixed within the harvester housing, thus only the change in flux density due to the moving magnets is considered. A is the area enclosed by the coil windings and is implemented as a vector in Matlab in which the elements of the vector represent the area of windings with different radii. N is the number of total windings of a coil. From Eq. 2.8, k is defined as Eq. 2.9.

$$k = -\frac{dB}{dx} A \cdot N \tag{2.9}$$

Looking back at the architectures it becomes evident, that the gradient ${}^{dB}/_{dx}$ is not constant over x because the mean flux density within a coil changes with the position x. Therefore the dependence of k on the position x of the magnets relative to the coils is taken into account in this work.

It was previously shown, that B_{mean} is computed for each position of the coil relative to the magnets. The gradient ${}^{dB}/_{dx}$ is computed using the fixed steps of $dx = d_{Cu}$, i.e. the coil wire diameter (which is used as the spacing between windings). While Eq. 2.9 is a scalar function to improve the understanding, in the model both B and A are vectors. It is the sum of all elements of $B_{mean,i} \cdot A_i$ (for every winding of radius r_i), i.e. all contributions to the induced voltage, that is used to compute the coupling curve $k(x)$.

The area of the coil windings is calculated as follows, starting from the largest, outermost winding down to the smallest winding. Equation 2.10 represents cylindrical coils.

$$A_i = m \cdot \pi \left(R_{max} - (i-1) \cdot d_{Cu}\right)^2, \; i = 1..n+1 \tag{2.10}$$

where R_{max} is the radius of the outermost winding, n is the number of windings in the radial direction and m is the number of windings along the direction of motion, i.e. with constant radius (perpendicular to the cylindrical cross-section) so that $m \cdot n = N$. Schematics explaining all geometrical parameters used in this work are given in Fig. 2.8. The values of n and m are calculated under the assumption of a regular winding process, by determining N first. This is achieved by multiplying the cross-sectional area of the coil with the copper fill factor and dividing the result by the cross-sectional area of the copper wire according to Eq. 2.11.

$$N = \frac{4 \cdot l_{coil} \cdot (R_{max} - R_{min}) \cdot k_{Cu}}{\pi \cdot d_{Cu}^2} \tag{2.11}$$

where R_{min} is the radius of the innermost winding, l_{coil} is the coil height in the case of cylindrical architectures and k_{Cu} is the copper fill factor, a value from standard wire parameter tables [6], which describes the ideal percentage of a coil cross-section that can actually be filled with copper for a given wire diameter. The number of windings in the radial direction of the coil n is then calculated as the rounded result of the expression $\sqrt{{}^{N}/_{l_{coil}} \cdot (R_{max} - R_{min})}$. The remaining parameter m is the rounded down result of $m = {}^{N}/_{n}$.

The area enclosed by the windings of a rectangular coil is calculated according to Eq. 2.12.

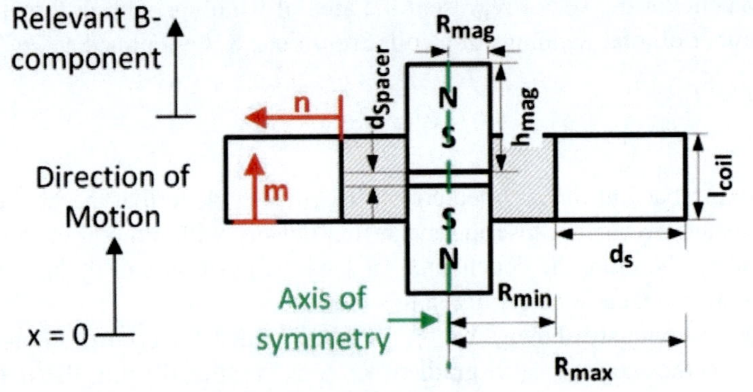

(a) Cylindrical architectures

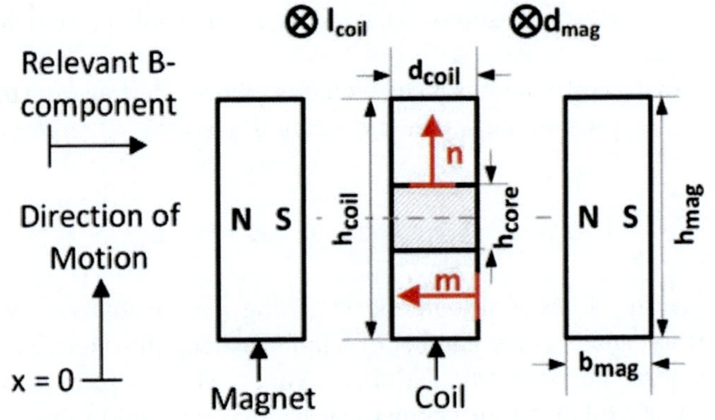

(b) rectangular architectures (rectangular coil shape)

Fig. 2.8 Schematics showing all relevant geometrical parameters of the two HAC types used in this work. The direction of magnet motion as well as the direction of the relevant component of magnetic flux density are also shown. Figures from [9]

$$A_{i+1} = m \cdot (h_{coil} - 2i \cdot d_{Cu}) \cdot (d_{coil} - 2i \cdot d_{Cu}) \, , \; i = 0..n-1 \qquad (2.12)$$

The parameters h_{coil} and l_{coil} describe the edges of the rectangular coil cross-section, with h_{coil} being located along the direction of motion as shown in Fig. 2.8. The thickness of the coil is named d_{coil} and h_{core} is the size of the air gap in the middle of the coil, which is non-zero due to the winding process. Matlab always starts vectors and matrices at index 1, unlike most other programming languages. Equations 2.10 and 2.12 are designed accordingly.

The above steps are repeated for every position along the motion direction at fixed steps of d_{Cu} leading to a coupling curve that spans the complete motion range. To this point, the calculations were performed as if the coils were moving with respect to the magnetic field. This is equivalent to fixed coils and moving magnets, thus the coupling curve can now be implemented as a look-up table within the electrical model in Simulink to determine U_{ind}, F_{el} and the power output of the different harvester architectures.

2.6 Induced Voltage, Electrical Damping Force and Power Output

In order to compute the induced voltage according to Eq. 2.8, the magnet's motion speed signal $\dot{x}$ is taken from the output port of INT1 in Fig. 2.3 and routed to the electrical model shown in Fig. 2.9. The actual motion of the the magnet is defined within a range according to Eq. 2.13.

$$-\frac{x_{max} - d_{gen}}{2} \leq x \leq \frac{x_{max} - d_{gen}}{2} \tag{2.13}$$

where the length of the magnet stack d_{gen} is subtracted from the full motion space within the harvester housing. The range of the look-up table is defined accordingly.

The blue block named "distInnen" is an array of the coil positions within the harvester housing. The simulations are set up to allow for a varying number of coils up to a maximum of 20. The spacing between the coils is equal to $h_{mag} + d_{Spacer}$ for cylindrical architectures and h_{mag} of the rectangular ones. Simulink however, does not support signals of flexible size. Therefore "distInnen" is always defined as a 20-component array, leading to a multiplexed 20-component signal in Simulink. Each component contains the center position of a single coil along x (with coils placed symmetrically around the center of the harvester housing at $x = 0$). If fewer than 20 coils are required for a simulation, the unused components of "distInnen" are set to a distance of 10 m and their coil resistance is neglected in further calculations. See Eq. 2.14, where n_{coils} is the number of "active" coils to be simulated, a number that can be smaller than the maximum of 20 coils that the model is prepared for:

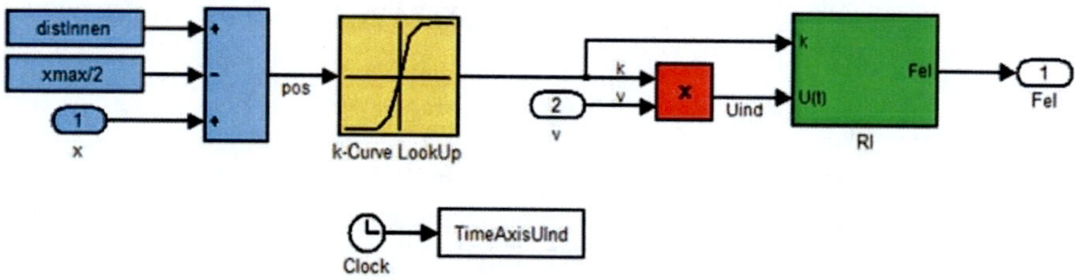

Fig. 2.9 Electromagnetic model components in the Simulink environment. The coil position is calculated in the blue blocks, the coupling coefficient look-up table is marked in yellow, the red block calculates U_{ind} and the green block calculates the power output and the electrical damping force which is required as a feedback input to the mechanical subsystem. Figure from [10]

$$distInnen(i) = \begin{cases} -\frac{n_{coils}-1}{2} \cdot (h_{mag} + d_{Spacer}) + (i-1) \cdot (h_{mag} + d_{Spacer}) & i = 1..n_{coils} \\ 10 & i = (n_{coils}+1)..20 \end{cases} \quad (2.14)$$

At the hypothetical distance of 10 m the coils do not experience a magnetic field (the look-up table of the coupling curve extrapolates to zero outside its active range) and are thus effectively neglected without reducing the flexibility of the model.

For each time step in the simulation and for every single coil, a value of k is extracted from the look-up table. This multiplexed signal is multiplied with $v = \dot{x}$ to calculate the induced voltage for each coil. A demultiplexer block is used within the green block in Fig. 2.9 to separate the voltage signals and add them up to generate the total induced voltage U_{ind}.

The electrical damping force F_{el} is computed according to Eq. 2.15.

$$F_{el} = i \cdot k \quad (2.15)$$

where i is the electrical current in the coils. To determine i from the previously calculated voltage, the resistance of the coils is also required. The inner resistance R_{in} of a cylindrical coil is calculated using Eq. 2.16.

$$R_{in} = 2\pi \cdot r_{mean} \cdot \varrho \cdot N \quad (2.16)$$

where r_{mean} is the mean winding radius for that coil and ϱ is the specific resistance of the copper wire measured in Ohm/m (ϱ is a function of the wire diameter and can be looked up in tables for the "IEC 60317" standard for air coils, e.g. in [6]). For wire with $d_{Cu} = 0.1\,mm$ the nominal specific resistance is $\varrho = 2.176\,\Omega/m$.

In the case of rectangular coils, the total wire length used of the coil is determined first (see Eq. 2.17).

$$l_{wire} = m \cdot \sum_{i=0}^{n-1} 2 \cdot (h_{coil} - 2i \cdot d_{Cu}) + 2 \cdot (d_{coil} - 2i \cdot d_{Cu}) \quad (2.17)$$

The resistance of the coil can then be calculated using Eq. 2.18.

$$R_{in} = \varrho \cdot l_{wire} \quad (2.18)$$

In this work, the electrical load applied to the harvester in simulations is always assumed to be a purely ohmic load. The inductance of the coils L is neglected because the complex impedance $j\omega L$ is small compared to the ohmic resistance of the coils. Due to the application at hand, i.e. energy harvesting from human motion, the excitation is non-harmonic and chaotic. The considered harvester architectures therefore do not contain springs and it is assumed that at the low step frequencies of the human gait of 1 Hz, no resonant effects occur. Consequently $\omega = 2 \cdot \pi \cdot f$ is very small at $f = 1\ Hz$ and the effect of the inductance L is not considered any further. The load R_l equals the sum of the total coil resistance following Eq. 2.19.

$$R_l = n_{coils} \cdot R_{in} \tag{2.19}$$

The load resistance is not set as $R_l = R_{in} + k_s^2/d_m$, as is the case with resonant harvesting devices where k_s is the spring coefficient and d_m is the mechanical damping. This is based on the aforementioned assumption, that no resonant effects occur due to the chaotic nature of the human gait as an excitation source.

At this point the voltage across the load is determined using a voltage divider and Simulink writes the load voltage U_l for each timestep of the simulation into an array in the Matlab workspace. Because Simulink is capable of variable-timestep simulations (which may potentially reduce the simulation runtime significantly in comparison to fixed-timestep simulations), the timesteps are written into a separate array in the Matlab workspace by the white blocks at the bottom of Fig. 2.9. The load voltage is squared one element at a time and divided by the scalar load resistance to give the instantaneous power output of the system for the simulation runtime. The average power output can now be computed directly.

2.7 Model Simplifications and Parameter Uncertainties

The previous sections described the modeling process. At this point, it has to be kept in mind, that the above system model can only be a simplified depiction of a highly complex problem. The acceleration data used in the simulations for instance was recorded along the plane of the foot. As the foot follows an arc during leg swing, the gravitational force acts on the system in various degrees. The angle between the foot and the floor however, is not considered in the system model.

An additional difficulty arises from the uncertainty associated with some of the parameters used in the system model. Parameters that are very hard to determine experimentally include the coefficients of friction and the coefficient of restitution. It was mentioned, that the coefficients of friction were fixed across all simulations performed in this work. Table 2.1 is used to explain this decision. It shows the normalized power output for HAC1 across a range of different friction coefficients. The data therein is normalized to $P_N = 1$ at $\mu_k = 0.2$ and $\mu_s = 0.3$. In the first six rows of Table 2.1 the difference between μ_k and μ_s is 0.1, while it is increased to 0.2 in the remaining rows. These values were chosen to cover the approximate range of friction coefficients for friction between plastic and metal bodies. The reader may notice, that the influence of the friction coefficients on the power output is comparatively small. HAC1 is used exemplarily to represent all considered architectures.

Table 2.1 Normalized power output as a function of the coefficients of friction in the case of HAC1. The motion data from test person "1" at 4 km/h is used as input data. Results normalized to $P_N = 1$ at $\mu_k = 0.2$ and $\mu_s = 0.3$. The influence of the coefficients of friction on the power output is relatively small. Table from [9]

Coefficient of restitution	Kinetic friction coefficient	Static friction coefficient	Normalized power output
k_i	μ_k	μ_s	P_N
0.6	0.15	0.25	1.01
0.6	0.2	0.3	1.00
0.6	0.25	0.35	0.98
0.6	0.3	0.4	0.97
0.6	0.35	0.45	0.96
0.6	0.4	0.5	0.94
0.6	0.15	0.35	1.01
0.6	0.2	0.4	1.00
0.6	0.25	0.45	0.99
0.6	0.3	0.5	0.97
0.6	0.35	0.55	0.96

Table 2.2 Normalized power output as a function of the coefficient of restitution in the case of HAC1. The motion data from test person "1" at 4 km/h and 6 km/h is used as input data. Results normalized to $P_N = 1$ at $k_i = 0.6$ respectively. Table from [9]

Coefficient of restitution	Kinetic friction coefficient	Static friction coefficient	Normalized power output (4 km/h)	Normalized power output (6 km/h)
k_i	μ_k	μ_s	$P_{N,4km/h}$	$P_{N,6km/h}$
0.5	0.2	0.3	0.97	0.96
0.6	0.2	0.3	1.00	1.00
0.7	0.2	0.3	1.03	1.05
0.8	0.2	0.3	1.09	1.12
0.9	0.2	0.3	1.14	1.20

The coefficient of restitution k_i on the other hand has a significant effect on the power output. Table 2.2 shows the normalized power output in the case of HAC1. The data is normalized to $P_N = 1$ at $k_i = 0.6$ for the motion speeds of 4 km/h and 6 km/h. A coefficient of restitution of $k_i = 1.0$ represents an ideal reflection without loss of kinetic energy and $k_i = 0$ represents the worst case, where all kinetic energy is lost and reflection occurs at the hard stop. The power output is rounded to three significant digits. The impacts of the magnets on the housing were modeled as hardstops with a fixed coefficient of restitution. This is a simplification of the real-world conditions in which the rubber end stops are compressed to a different degree depending on the mass of the moving magnet stack and its acceleration/speed upon impact. Table 2.2 indicates that k_i may have a larger effect on the power output as the walking speed increases, i.e. as the accelerations of the system increase and the magnet stack moves faster within the housing.

This is an important issue to be investigated in future work, where the speed-dependency of the coefficient of restitution could be analyzed using a dedicated measurement setup. A range of materials can consequently be analyzed to determine a suitable end stop for this type of device. This however, is beyond the scope of this work.

2.8 Discussion

The above sections explained one possible implementation of how different inductive HACs can be modeled in the Matlab/Simulink environment. The model is modeled in such a manner, that almost every aspect of it is parameterized and flexible. In the following chapter, the established model is used to optimize the design parameters of each architecture in order to find a configuration with maximum power output under consideration of the highly restricted space that is available in the sole of a shoe. A simple change in the external acceleration data as well as the parameters that interface to the real world, e.g. the maximum dimensions of the device and the motion range, allows the use of this model for other walking conditions or other applications, that differ entirely from energy harvesting at the human body.

Difficulties with the model may arise, e.g. when acceleration data with significant noise around zero is used. This leads to a high number of zero-crossing in the Simulink model which increases the simulation time and in the worst case may cause it to abort the simulation. If such a case occurs, the input acceleration data needs to be pre-conditioned using filters or a smoothing algorithm.

Another problem may be caused by the friction model, where the used parameters are not measured, but assumed values. Approaches for parameter-fitting using extensive experimental datasets have been presented in literature [7, 8]. The generation of the required experimental data as well as the application of the mentioned methods is considered as a follow-up to this work and is expected to provide more confidence in the parameters.

References

1. W.S. Levine, *The Control Handbook*. The Electrical Engineering Handbook Series (CRC Press and IEEE Press, Boca Raton, FL, New York, NY, 1996)
2. MathWorks, Translational friction (2015), http://de.mathworks.com/help/simulink/examples/friction-model-with-hard-stops.html. Accessed 20 Oct. 2015
3. F. Altpeter, *Friction Modeling, Identification and Compensation*. Ph.D. thesis, École Polytechnique Fédérale de Lausanne, Lausanne (1999)
4. H.A. Gall, Zur simulation von haftreibung und mechanischen anschlägen. ASIM Nachrichten **1**, 4–9 (2001)
5. D. Meeker, Finite element method magnetics (femm) (2017), http://www.femm.info/wiki/homepage. Accessed 03 March 2017
6. Elektrisola, Iec 60317 standard—technical data for self bonding copper wire (2017), http://www.elektrisola.com/de/backlackdraht/technische-daten-nach-durchmessern/iec-60317-ueberblick.html. Accessed 03 March 2017
7. P.L. Green, K. Worden, N.D. Sims, On the identification and modelling of friction in a randomly excited energy harvester. J. Sound Vib. **332**(19), 4696–4708 (2013)
8. P.L. Green, K. Worden, K. Atallah, N.D. Sims, The effect of duffing-type non-linearities and coulomb damping on the response of an energy harvester to random excitations. J. Intell. Mater. Syst. Struct. **23**(18), 2039–2054 (2012)
9. K. Ylli, Energy harvesting from the swing motion of the human leg (2019)
10. K. Ylli, D. Hoffmann, A. Willmann, B. Folkmer, Y. Manoli, Human motion energy harvesting: numerical analysis of electromagnetic swing-excited structures. Smart Mater. Struct. **25**(9), 095014 (2016)

Chapter 3
Geometrical Parameter Optimization

3.1 Optimization Procedure

The previous chapter described the development of a system model which uses measured acceleration data to simulate the motion of the magnet stack within an inductive EH device. From the magnet motion relative to the coils, the induced voltage across an ohmic load is determined and subsequently the average power output is calculated.

Every simulation represents one specific set of geometrical parameters (GP), e.g. magnet height and number of magnets used. It can be assumed that a set of optimal geometrical parameters exists for each particular application and its respective building space. In this work harvesting devices are developed, which are integrated into the shoe sole and use the human leg motion as an energy source.

The device dimensions are restricted by the application at hand. Through several iterative steps with prior experiments, the housing was strengthened, so it could carry the weight of the user without being damaged. The height of the devices is limited to 15.75 mm, while the length and width are limited to 77 mm and 41.5 mm respectively (see Sect. 1.3 for a detailed explanation of the size constraints). The following chapters show the fabricated devices which all use 3D-printed housings adapted for the harvester architecture (HAC). The walls of the housing are approximately 2 mm thick, resulting in the maximum possible dimensions of the active transducer volume of 71 mm in length, 12.5 mm in height and 37.5 mm in width. These boundary conditions are identical for all architectures.

Given these size constraints, a parameter optimization is performed, in order to find the best possible magnet and coil setup. Each combination of parameters is called a "design". The optimization process initially requires a design of experiments (DOE). In this work the DOE consists of a total of 50 designs that are implemented as a "Sobol sequence" which represents a near-uniform yet quasi-random distribution within a multi-dimensional parameter space [1]. The average power output is the target function and the external acceleration data that serves as the system input

K. Ylli and Y. Manoli, *Energy Harvesting for Wearable Sensor Systems*,
Springer Series in Advanced Microelectronics 62,
https://doi.org/10.1007/978-981-33-4448-8_3

Table 3.1 Parameter boundaries for the geometrical optimization of HAC1. Table from [5]

Parameter	Description	Lower bound	Upper bound
h_{mag}	Magnet height	0.5 mm	25 mm
R_{mag}	Outer magnet radius	1 mm	4.5 mm
r_{mag}	Inner magnet radius	0 mm	4 mm
l_{coil}	Coil height	1 mm	25 mm
d_{coil}	Coil wall thickness	0.5 mm	4 mm
d_{Spacer}	Spacer thickness	0 mm	4 mm
n_{mag}	Number of magnets	2	20
n_{coils}	Number of coils	1	20

remains the same across all simulations. Throughout this chapter, the acceleration data recorded for runner "1" on a treadmill at 4 km/h is used in the simulations.

Within the optimization software "modeFrontier" [2] the genetic algorithm "MOGA-2" is used. Each optimization step generates a new and modified set of 50 designs and thus forms a new "generation". The algorithm is set to use a total of 100 generations to find the optimum design. However, the number of designs calculated is smaller than the possible maximum of 5000, as many designs do not fulfill the constraints specific to each architecture and repeated designs may be calculated during the optimization.

The procedure of the geometrical optimization process is exemplarily explained for HAC1 and is the same for the remaining architectures. Each architecture is set up to be fully flexible. With respect to HAC1 the magnets can either be (hollow) ring magnets or full cylindrical magnets. The number of magnets and coils is variable. The used parameter space is given in Table 3.1. For HAC1 the magnets are axially polarized and placed in opposing orientation as shown in Fig. 1.14. The inner diameter of the coils, whenever it is not directly related to the magnet size, is defined to be 1 mm (as is the case in HAC3-HAC6).

The constraints that limit any possible design to a feasible size for enabling the integration into a shoe are given in Eqs. 3.1–3.6 using the parameter abbreviations from Table 3.1. Constraint (3.1) limits the height of the magnet stack in such manner that at least 10 mm of motion space are left within the housing. This is an empirical value determined from previous simulations and experiments, where a smaller motion space had a strongly negative effect on the power output. Constraint (3.2) limits the radius of the magnet-coil setup to 6.25 mm, thus resulting in a total height of 12.5 mm. The 0.75 mm gap between magnets and coils represents the plastic channel through which the magnets move. Constraint (3.3) forces the outer magnet radius to be at least 0.1 mm larger than the inner magnet radius. Constraint (3.4) sets the spacer thickness to a certain minimal value, in order to avoid errors from the FEM simulation. Constraint (3.5) limits the coils to the magnets in terms of height, so that the coils can be placed at equal intervals between magnets and generate a voltage synchronously. The coils have to be slightly smaller than the magnets, so that they can

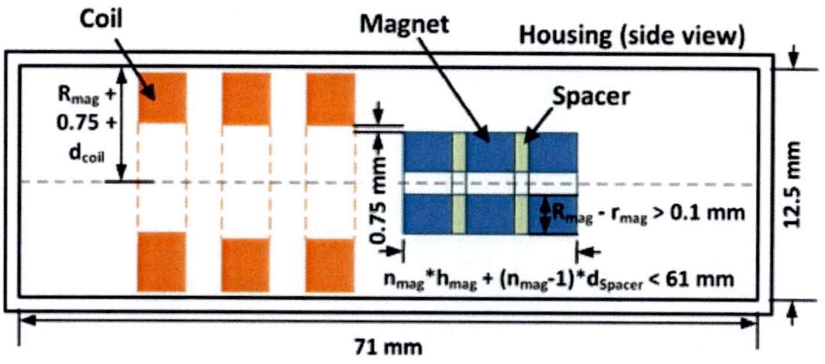

Fig. 3.1 Cross-sectional schematic in side view that explains three optimization constraints exemplarily for HAC1. Similar constraints exist for all harvester architectures. Figure from [4]

be fixed within the housing. The last constraint (3.6) limits the total length occupied by the coils within the housing (see Fig. 3.1).

$$71\,\text{mm} - n_{mag} \cdot h_{mag} - \left(n_{mag} - 1\right) \cdot d_{Spacer} - 10\,\text{mm} > 0\,\text{mm} \quad (3.1)$$

$$R_{mag} + 0.75\,\text{mm} + d_{coil} - 6.25\,\text{mm} < 0\,\text{mm} \quad (3.2)$$

$$R_{mag} - r_{mag} > 0.1\,\text{mm} \quad (3.3)$$

$$d_{Spacer} - 2 \cdot 0.01 \cdot R_{mag} > 0\,\text{mm} \quad (3.4)$$

$$l_{coil} - h_{mag} - 0.5\,\text{mm} < 0\,\text{mm} \quad (3.5)$$

$$n_{coils} \cdot l_{coil} < 71\,\text{mm} \quad (3.6)$$

3.2 Optimization Results

3.2.1 *HAC1—Axially Polarized Cylindrical Magnets*

The optimization results are summarized in a parallel coordinate graph as shown in Fig. 3.2. It shows all calculated designs with their corresponding parameters in one diagram. Each column represents a parameter and its boundaries from the lower to

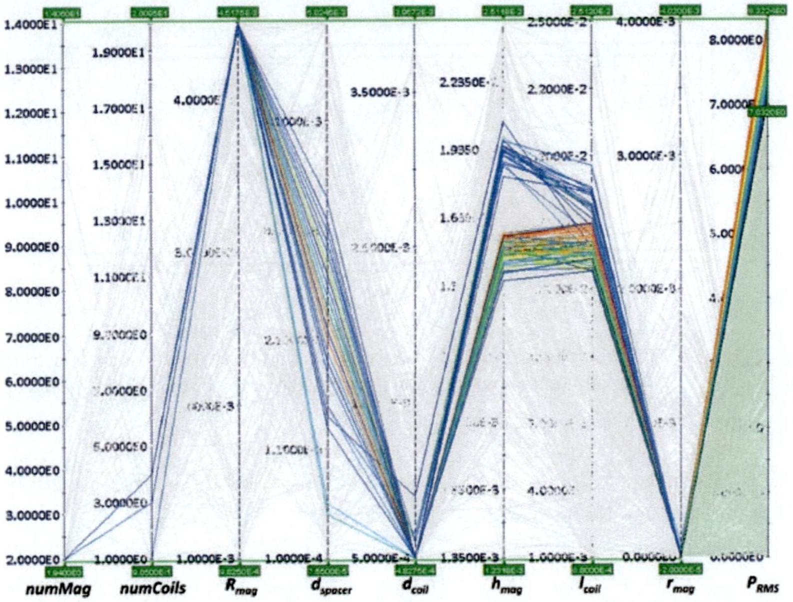

Fig. 3.2 Parallel Coordinate Chart that shows all calculated designs of HAC1. Each column represents one parameter. Right-most is the average power output where a filter has been applied, highlighting the best designs. A clear tendency is visible for two parameters: R_{mag} has to be as large as possible, while r_{mag} has to be zero which means that cylindrical magnets are more feasible than ring magnets. Figure from [5]

the upper limit. The right-most column is the target variable, in this case the average power output. Sliders can be applied within the software to filter the parameters, e.g. only the designs within a defined power output band are shown. This enables the fast determination of tendencies of the parameters. In the particular case of HAC1 (see Fig. 1.14) it is immediately noticeable that, as expected, cylindrical magnets have to be used instead of ring magnets as the tendency for R_{mag} is to have the maximal value within its boundaries, while r_{mag} is zero. Additionally, the number of magnets and coils tends to be small, while the magnet and coil height is relatively large. The magnet and coil height are quite similar across all filtered designs. This gives the opportunity to make best use of the available spacing between magnets.

The parameters for the best design and the corresponding maximum average power output are listed in Table 3.2. At this point it has to be kept in mind, that HAC1 is a cylindrical structure, i.e. the optimum design of the total structure is 12.5 mm in diameter (equal to the height available in the housing). This in turn means that two of these magnet-coil setups can be placed in parallel within the housing which is 37.5 mm wide and effectively double the power output, unless the magnets setup influence each other due to attractive forces. Although three magnet-coil setups would

Table 3.2 Geometrical Parameters and average power output of the best design found for HAC1. All lengths are given in mm. Table from [5]

R_{mag}	r_{mag}	h_{mag}	l_{coil}	d_{coil}	d_{Spacer}	n_{mag}	n_{coils}	P_{RMS}
4.5	0	15.55	15.95	0.5	2.2	2	4	8.28 mW

fit within the housing, this is not feasible due to the magnetic interaction between the magnet stacks. Although separated by the coils, the magnet stacks would be forced to the channel walls, which creates large friction. In Chap. 4 it is shown, that there is no general answer to whether two magnet setups can be placed within the same housing, as it strongly depends on the chosen magnets.

3.2.2 HAC3—Coils Within Axially Polarized Ring Magnets

HAC3 differs from the previous architecture HAC1 in terms of the placement of the coils. Whereas in HAC1 the coils surround the magnets, they are now placed inside the magnets as shown in Figs. 1.14 and 3.3b. Therefore, the components of the flux density that are perpendicular to the coils' surface (and contribute to the induced voltage) are different.

The geometrical parameters of HAC3 (Table 3.3) and their respective boundaries are defined as follows. The key difference from the previous architectures is that the magnet dimensions are defined indirectly through the coil dimensions and magnet thickness in such manner that the available radius of 6.25 mm is used fully. Addi-

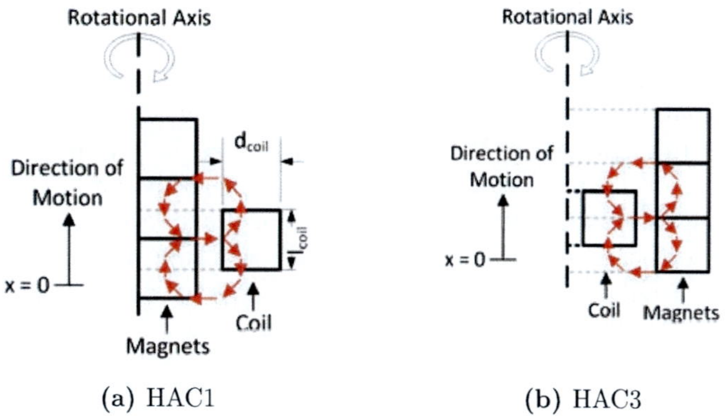

Fig. 3.3 **a** Schematic of HAC1 with coils surrounding the axial magnets. The magnetic flux is shown in red. **b** Schematic of HAC3. The coils' enclosed area is crossed by different flux paths than in HAC1, where the flux path may close within the coil. Figures from [5]

Table 3.3 Parameter boundaries for the geometrical optimization of HAC3. Table from [5]

Parameter	Description	Lower bound	Upper bound
h_{mag}	Magnet height	0.5 mm	25 mm
d_{mag}	Magnet wall thickness	0.5 mm	4.5 mm
d_{iron}	Back iron thickness	0 mm	3 mm
l_{coil}	Coil height	1 mm	25 mm
d_{coil}	Coil wall thickness	1 mm	4 mm
d_{Spacer}	Spacer thickness	0 mm	5 mm
n_{mag}	Number of magnets	2	20
n_{coils}	Number of coils	1	20

Table 3.4 Geometrical Parameters and average power output of the best design found for HAC3. All lengths are given in mm. Table from [5]

d_{mag}	d_{iron}	h_{mag}	l_{coil}	d_{coil}	d_{Spacer}	n_{mag}	n_{coils}	P_{RMS}
2.9	0	3.4	1.65	2.05	1.0	9	12	7.05 mW

tionally a back iron cylinder may be placed around the magnet stack to guide the magnetic flux as shown in Fig. 1.14.

$$0.5\,\text{mm} + d_{coil} + 0.75\,\text{mm} + d_{mag} + d_{iron} - 6.25\,\text{mm} < 0\,\text{mm} \tag{3.7}$$

$$d_{Spacer} - 2 \cdot 0.01 \cdot \left(0.5\,\text{mm} + d_{coil} + 0.75\,\text{mm} + d_{mag}\right) > 0\,\text{mm} \tag{3.8}$$

For this architecture the constraints (3.1), (3.5) and (3.6) are valid. The additional constraint (3.7) describes the system radius which is now calculated using the magnet and back iron thickness. Constraint (3.8) is adapted from (3.4) and expresses the outer magnet radius in terms of the magnet thickness.

The best design for HAC3 is listed in Table 3.4. A back iron cylinder does not offer an advantage as indicated by the optimization result of $d_{iron} = 0$, instead the magnet rings are comparatively thick. The height of both magnets and coils is small and their number is comparatively large, leading to a large number of magnetic flux changes in a coil during one motion period. The average power output of 7.05 mW is slightly lower than for the previous architecture.

3.2.3 *HAC5—Rectangular Closed Magnetic Circuit*

HAC5 is one of the two rectangular architectures considered in this work (see Fig. 1.14). Flat magnets are used in a closed magnetic circuit, which span the full inner

Table 3.5 Parameter boundaries for the geometrical optimization of HAC5. Table from [5]

Parameter	Description	Lower Bound	Upper Bound
h_{mag}	Magnet height	1 mm	25 mm
b_{mag}	Magnet thickness	0.5 mm	5 mm
d_{coil}	Coil height	1 mm	8 mm
h_{coil}	Coil length	2 mm	10 mm
d_{iron}	Back iron thickness	0 mm	5 mm
n_{mag}	Number of magnets	2	20
n_{coils}	Number of coils	1	20

width of the housing (37.5 mm). HAC5 can thus make a better use of the available volume within a shoe sole than can HAC1 and HAC3. The required geometrical parameters and their boundary conditions are listed in Table 3.5. In order to make efficient use of the available space, the coils are implemented as rectangular coils in the system model. The coil length h_{coil} and coil height d_{coil} (according to Fig. 2.8) are variable, while the width of the coil is set to 37.5 mm which corresponds to the space available within the harvester housing. The air gap h_{core} within the coil center is 1 mm. The magnets are placed in such a way that several flux changes can occur within the coils along the direction of motion.

HAC5 is accompanied by a new set of constraints given by Eqs. 3.9 to 3.12. Constraint (3.9) sets the magnet stack to be at least 10 mm shorter than the housing to allow motion. Constraint (3.10) limits the height of the device including the two air gaps of 0.5 mm between magnets and coils. Constraint (3.11) limits the coil rows in terms of their length along the direction of motion to the housing dimensions. The last constraint (3.12) limits every coil's length to be smaller than the magnet height, so that each coil experiences a full flux change between magnets without interference from the neighboring flux change, when many magnets are used. Some space between the coils is also required during fabrication, to fix them within the housing.

$$71\,\text{mm} - n_{mag} \cdot h_{mag} - 10\,\text{mm} > 0\,\text{mm} \tag{3.9}$$

$$2 \cdot b_{mag} + 2 \cdot d_{iron} + l_{coil} + 1\,\text{mm} < 12.5\,\text{mm} \tag{3.10}$$

$$h_{coil} \cdot n_{coils} < 71\,\text{mm} \tag{3.11}$$

$$h_{coil} - h_{mag} < 0\,\text{mm} \tag{3.12}$$

Table 3.6 Geometrical Parameters and average power output of the best design found for HAC5. All lengths are given in mm. Table from [5]

b_{mag}	d_{iron}	h_{mag}	d_{coil}	h_{coil}	n_{mag}	n_{coils}	P_{RMS}
1.25	3.8	16.6	1.35	9.8	4	5	42.66 mW

The best design for HAC5 can be found in Table 3.6. Very flat coils and magnets are the result of the optimization with 1.35 mm and 1.25 mm respectively. The back iron is rather thick compared to the magnets, which may be explained by an improved flux guidance. The estimated average output power of 42.66 mW outperforms the previous architecture by a wide margin.

3.2.4 HAC6—Rectangular Double Halbach Array

This architecture differs from HAC5 only in the orientation of the magnets, which are oriented to form a two-sided Halbach Array (called a double Halbach Array) as shown in Fig. 1.14. The geometrical parameters from HAC5 are still valid with the exception of the back iron which is not used in HAC6. The Halbach Array is inherently shielded due to the unique orientation of neighboring magnets which guide the magnetic flux towards the inside of the magnet arrangement [3]. Since four different magnet orientations are required for HAC6, two of them perpendicular to the direction of motion and two aligned with it, a second magnet height $h_{mag,2}$ is introduced with the same boundaries as h_{mag} and represents the latter two orientations. Constraint (3.11) remains valid, while the other constraints from HAC5 have to be modified slightly as follows. The length of the magnet stack in (3.13) has to be adapted to allow for the two different magnet heights. The system height in (3.14) no longer contains the back iron and constraint (3.15) limits the coil height to the sum of the two magnet heights due to the flux paths within the Halbach Array.

$$71\,\text{mm} - \lceil n_{mag}/2 \rceil \cdot h_{mag} - \lfloor n_{mag}/2 \rfloor \cdot h_{mag,2} - 10\,\text{mm} > 0\,\text{mm} \tag{3.13}$$

$$2 \cdot b_{mag} + l_{coil} + 1\,\text{mm} < 12.5\,\text{mm} \tag{3.14}$$

$$h_{coil} - h_{mag} - h_{mag,2} < 0\,\text{mm} \tag{3.15}$$

The best design for HAC6 as presented in Table 3.7 concludes the optimization results of the four investigated architectures. The average power output of 23.2 mW is lower than for HAC5 and the most notable parameter is the number of coils used. Only a single rectangular coil results in the optimization. This simulation result can be attributed to the high electromagnetic damping caused by the Halbach configuration

Table 3.7 Geometrical Parameters and average power output of the best design found for HAC6. All lengths are given in mm. Table from [5]

b_{mag}	h_{mag}	$h_{mag,2}$	d_{coil}	h_{coil}	n_{mag}	n_{coils}	P_{RMS}
5.0	14.8	13	1.35	9.1	6	1	23.2 mW

of the magnets. The magnetic flux is guided within the magnet arrangement in a way that a large flux change occurs in the coil [3] and consequently a large coupling coefficient is achieved. Considering Eqs. 2.8 and 2.15 one can see that the electrical damping force is proportional to the squared coupling coefficient $\left(F_{el} \sim k^2\right)$ and thus large electrical damping may already be achieved with a single coil. The large damping can hinder the motion of the magnet stack, which in turn leads to a decreased power output. An increased number of coils can lead to an additional decrease in the power output.

3.3 Discussion

This chapter presented the optimization of the geometrical parameters for HAC1-HAC6 based on the system model described in Chap. 2. The best set of geometrical parameters was presented for each architecture and based on the predicted average power output it is possible to choose the most suitable architecture. A summary is given in Table 3.8. For the cylindrical HAC1 and HAC3 the total power output $P_{RMS,total}$ is based on the fact that two magnet stacks may fit within the housing (i.e. the volume 77 mm in length, 41.5 mm in width and 15.75 mm in height mm^3). This means that the power output of a single magnet-coil-stack can potentially be doubled through parallelization within the housing, as long as the effect of attractive forces between magnet stacks is negligible. The total device volume including the housing is 50.33 cm^3.

The first observation is that the cylindrical architectures seem to generate less energy than the two rectangular ones. Even when the available width within the sole is considered, i.e. two cylindrical systems can be placed next to each other, the average power output of the doubled HAC1 is 16.56 mW as opposed to the 42.6 mW expected from HAC5. Considering the total device volume including the housing, the rectangular designs have an even greater advantage. The cylindrical systems attract each other magnetically, i.e. only one or two separate transducer stacks can be placed within a housing, although three would fit. Therefore a significant amount of volume is wasted unnecessarily. This is reflected in the comparatively poor power density (total volume) figures in Table 3.8.

An additional advantage of architectures with back iron is the magnetic shielding. As the magnetic field is guided through the iron parts, nearly no flux density crosses to the exterior of the device, which might be a useful feature if used in a different environment than the human body, where magnetic materials may be present.

Table 3.8 Summary of the optimized designs for architectures HAC1 to HAC6. In the case of cylindrical HACs, the "total" volume refers to a housing that is wide enough for two magnet stacks to be placed within it. Table modified from [5]

Architecture	HAC1	HAC3	HAC5	HAC6
Type	Cylindrical	Cylindrical	Rectangular	Rectangular
P_{RMS} (mW) (per magnet stack)	8.28	7.05	42.60	23.20
$P_{RMS,total}$ (mW) (two stacks)	16.56	14.10	42.60	23.20
Active volume (cm^3) (one stack)	8.96	8.96	34.22	34.22
Volume of housing (cm^3)	50.33	50.33	50.33	50.33
Power density (μ W/cm^3) (active volume)	924.1	786.8	1244.9	678.0
Power density (μ W/cm^3) (total housing volume)	329.0	280.2	846.4	461.0

Of further interest is the comparison of the two rectangular architectures. The Halbach Array in HAC6 uses a single flat coil to generate an average power output of 23.2 mW. It was previously mentioned that the electromagnetic coupling of the Halbach Array is very high due to the special magnet arrangement. The high coupling leads to a large electrical damping force F_{el} which has a negative effect on this indirectly excited system. HAC5 proves that high coupling does not necessarily imply optimal coupling as a high F_{el} reduces the magnet motion to such a degree that the power output can drop significantly.

Special attention needs to be payed to the FEM simulations of magnetic flux distribution. FEM simulations can be very sensitive to the simulation setup, thus a good configuration needs to be found and ideally needs to be kept constant across all simulations. An alternative for FEM would be the analytical solution of the magnetic flux distribution. Herewith the problem is that a closed-form solution exists only for a limited variation of magnet arrangements (generally not including back iron parts).

Generally speaking the magnet motion model employs several simplifications and one cannot expect a fully realistic representation of the actual motion. One example is the tilting of the foot during motion, which has not been considered in this work. It can however influence the simulation outcome, since the earth's gravity will nearly always contribute to the external accelerations at a certain angle (except at standstill). The friction model is one possible representation of a highly complex problem which neglects in-depth considerations like surface roughness etc.

In conclusion HAC5 was found to be the most suitable architecture within the size constraints posed by the integration into the shoe sole. A simulated power output of more than 42 mW with an ohmic load offers the opportunity to power body-worn systems directly and without the need for maintenance and battery replacement. In the following chapter the best designs will be engineered in CAD, fabricated and evaluated experimentally. It has to be kept in mind however, that an experimental setup cannot be built to fully match the found parameters as for example magnets are not readily available in step sizes of 50 μm as would be required here. Tolerances in the components' dimensions were not considered in this chapter. It is immediately clear that when using a coil wire with 100 μm thickness for example, it will not be possible to achieve a coil length of 15.95 mm accurately. Therefore, deviation from these theoretical results are expected and will be analyzed. The possibility of further mechanical improvements will be investigated.

References

1. I.M. Sobol', D. Asotsky, A. Kreinin, S. Kucherenko, Construction and comparison of high-dimensional sobol' generators. Wilmott **2011**(56), 64–79 (2011)
2. modeFrontier, modefrontier optimization environment (2017), http://www.esteco.com/modefrontier. Accessed 03 April 2017
3. D. Zhu, S. Beeby, J. Tudor, N. Harris, Vibration energy harvesting using the Halbach array. Smart Mater. Struct. **21**(7), 075020 (2012)
4. K. Ylli, Energy harvesting from the swing motion of the human leg (2019)
5. K. Ylli, D. Hoffmann, A. Willmann, B. Folkmer, Y. Manoli, Human motion energy harvesting: numerical analysis of electromagnetic swing-excited structures. Smart Mater. Struct. **25**(9), 095014 (2016)

Chapter 4
Experimental Evaluation of Fabricated Architectures

4.1 Introduction

This section gives the experimental results of the four investigated harvester architectures (HAC). They are comprised of two cylindrical HACs and two cuboid HACs as shown in Fig. 4.2. The numbers of the HAC correspond to the numbers used previously in this work to facilitate comparison. The fabricated devices and the obtained experimental results enable the revision and verification of the system models presented in Chap. 3 of this work.

4.2 Swing Harvester Evolution

During this work, several harvesting devices were fabricated. Before the detailed investigation of the four harvester architectures presented in this work, a number of devices were fabricated and characterized experimentally. They are not included in the detailed investigation of the four architectures, that is shown in the following sections and contains only the most recent generation of fabricated swing-type devices. However, those previous generations of devices were important in understanding this type of harvesting device, both in terms of modeling and fabrication. Figure 4.1 shows these previously fabricated devices that were first presented in [1, 2].

The main parameters for this previous generation of devices are given in Table 4.1. The harvester HH-V1(horizontal harvester) contains a single large cylindrical magnet and two coils [1]. MCG-V1 (multi-coil generator) and MCG-V7 are multi-coil devices.

They contain magnet stacks made of several ring magnets placed in opposing orientation with flux guiding spacers in between them. A thread rod of non-magnetic material is placed through all magnet rings to provide a larger contact surface for the used glue, that holds the magnets together against the opposing magnetic forces.

K. Ylli and Y. Manoli, *Energy Harvesting for Wearable Sensor Systems*,
Springer Series in Advanced Microelectronics 62,
https://doi.org/10.1007/978-981-33-4448-8_4

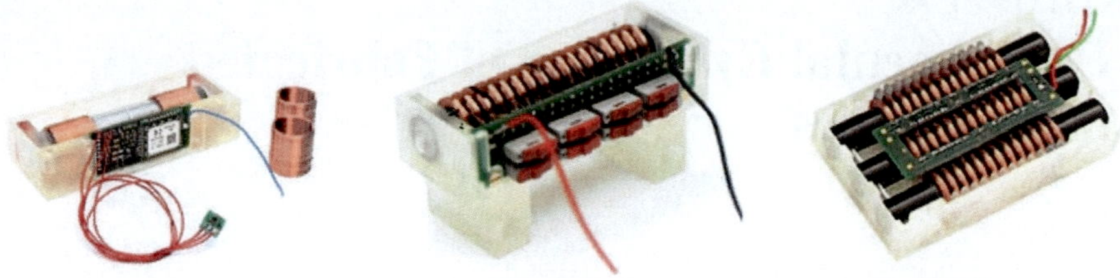

(a) HH-V1 (b) MCG-V1 (c) MCG-V7

Fig. 4.1 Pictures of devices that were previously fabricated during this work. These devices are not part of the detailed investigation shown in this work. Details on the devices are given in Table 4.1. Figure 4.1 (**a**) reproduced from [1]. Figure 4.1 (**b**) and (**c**) reproduced from [2]

Table 4.1 Key geometry parameters of the previous harvester iterations as shown in Fig. 4.1. Table from [9]

	Device (a): HH-V1	Device (b): MCG-V1	Device (c): MCG-V7
Size	$110 \times 40 \times 20\,mm^3$	$65 \times 25 \times 15\,mm^3$	$75 \times 41.5 \times 15\,mm^3$
Active size	$100 \times 19 \times 19\,mm^3$	$57 \times 12.5 \times 12.5\,mm^3$	$71 \times 37.5 \times 12.5\,mm^3$
Volume	$88\,cm^3$	$24.38\,cm^3$	$46.69\,cm^3$
Active volume	$36.1\,cm^3$	$8.91\,cm^3$	$33.28\,cm^3$
Coils	2	13	3 * 9
Magnets	1	14	3 * 17
Magnet stack length	50 mm	41 mm	52 mm
Weight (incl. housing)	141 g	44.5 g	82 g

Data on the power output of these three devices is given in Table 4.2. It can be seen that HH-V1 generates the largest power output during walking. This device however, is by far the largest and does not meet the size constraints defined in Sect. 1.3. MCG-V7 is a device where three magnet stacks are connected by braces at either end, in order to move simultaneously and not attract each other magnetically. The magnets are identical to MCG-V1 from [3] at 7 mm in outer diameter, 2.7 mm in inner diameter and 2 mm in height and so are the coils at 12.5 mm outer diameter, 8.5 mm inner diameter and a height of 1.2 mm. Due to the braces the number of coils that are used in MCG-V7 is reduced as is explained in more detail in [2].

The use of three parallel magnet stacks does not generate a threefold power output compared to MCG-V1 due to the reduced number of coils and the reduced motion freedom because of the longer magnet stack.

A version of MCG-V7 with only two separate magnet stacks in the outer positions of the housing (and without any interconnecting structure) was also briefly analyzed. It generated an average power output of 3.2 mW on a treadmill at 6 km/h for test person “1”. This is more than a threefold increase in power output over MCG-V1 and shows that it can be a feasible approach to place two cylindrical transducer setups within one housing. There was no noticeable attraction between the magnet

Table 4.2 Average power output of previously fabricated swing harvesting devices for test person "1". The measurements were performed on a treadmill using an ohmic load equal to the coil resistance of the devices. Table from [9]

	HH-V1		MCG-V1		MCG-V7	
Speed (km/h)	4	6	4	6	4	6
Power output (mW)	6.06	9.98	0.46	0.84	1.08	1.81
Power density ($\mu W/cm^3$) (active volume)	68.9 (167.9)	113.4 (276.3)	18.9 (51.7)	34.5 (94.3)	23.2 (32.5)	38.8 (54.4)
Power density ($\mu W/g$) (weight incl. housing)	43.0	70.8	10.3	18.9	13.2	22.1

stacks, which could hinder the magnet motion. As is shown in Sect. 4.7, however, this approach depends strongly on the used magnets and has to be evaluated for each transducer setup separately.

4.3 Fabrication

In Chap. 3 a geometrical parameter optimization was performed, in order to find the best magnet and coil setup for each of the four HAC under investigation. The volume, i.e. the available space within the 3D-printed housing is identical across all architectures. The devices are fabricated as closely as possible to the geometrical parameters that resulted from the parameter optimization. The magnets were not commercially available in the desired dimensions. For each HAC, a number of different available magnets were selected, that are close to the desired dimensions.

New simulations were performed with these dimensions and the magnets that produced the largest power output in the simulations, i.e. that most closely matched the optimized maximum output from the system simulations, were chosen for the experimental setup. These new simulations based on commercially available magnets are required in order to guarantee, that the coils and steel spacers match the chosen magnets. This explains the difference in the geometrical parameters between experiment and optimization, e.g. where a steel spacer with a thickness of 1.4 mm is used in the experiment for HAC1, instead of the thickness of 2.2 mm that was determined by the optimization (see 3.2.1).

All magnetic structures are placed within 3D-printed housings designed for each HAC. The outer housing size is 77 mm in length, 41.5 mm in width and 15.75 mm in height as defined in Sect. 1.3. The space within the housing is 71 mm in length, 37.5 in width and 12.5 mm in height. Rubber end stops effectively reduce the motion range to 67–69 mm. Multiple coils are used across all HACs and in each case they are placed with the same spacing as the magnets, enabling a series connection of the coils and the generation of one synchronous voltage across them. The geometry

Table 4.3 Key geometry parameters of the fabricated devices for HAC1 and HAC3. Table from [9]

Parameter	HAC1	HAC3
Magnet height	12 mm	4 mm
Magnet outer diameter	9 mm	12 mm
Magnet inner diameter	–	7.5 mm
Coil height	3.2 mm	1 mm
Coil inner diameter	10.5 mm	2 mm
Coil outer diameter	12.5 mm	6 mm
Number of magnets	3	8
Number of coils	4	9
Spacer height	1.4 mm	0.5 mm
Magnet stack length	38.8 mm	35.5 mm
Magnet stack weight	18.4 g	18.2 g
Weight (incl. housing)	62.5 g	63.5 g

Table 4.4 Key geometry parameters of the fabricated devices for HAC5 and HAC6. Table from [9]

Parameter	HAC5	HAC6
Magnet height	1 mm	2 mm
Magnet length	17 mm	18 mm
Magnet width	8 mm	4 mm
Coil height	1.2 mm	1.2 mm
Coil length	15.5 mm	11 mm
Coil width	6.5 mm	3.8 mm
Coil inner width	1.1 mm	1.1 mm
Number of magnets	20	38
Number of coils	16	12
Back steel height	4 mm	–
Magnet stack length	40 mm	38 mm
Magnet stack weight	95.8 g	40.4 g
Weight (incl. housing)	130 g	64.5 g

parameters are listed in Table 4.3 for the two cylindrical devices and in Table 4.4 for the two cuboid devices.

The cylindrical HACs contain magnet stacks formed by opposing cylindrical magnets glued together with steel spacers in between them. Therefore the magnetic flux is forced through the spacer. A 3D-printed gluing aid is used to align the magnets and spacers as precisely as possible. The surfaces of the magnets are roughened slightly using sandpaper and cleaned using isopropanol. A thin layer of the two-

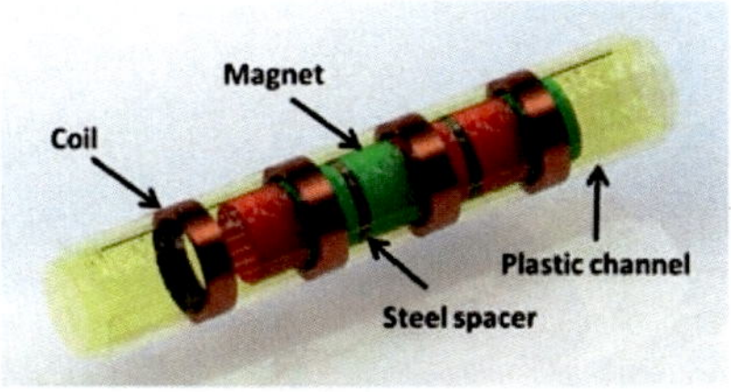

(a) HAC1

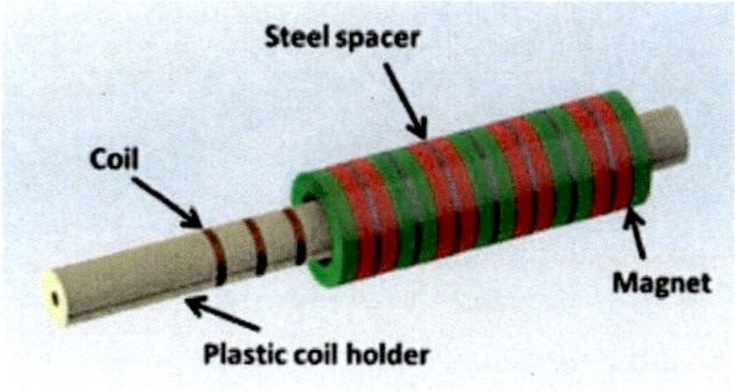

(b) HAC3

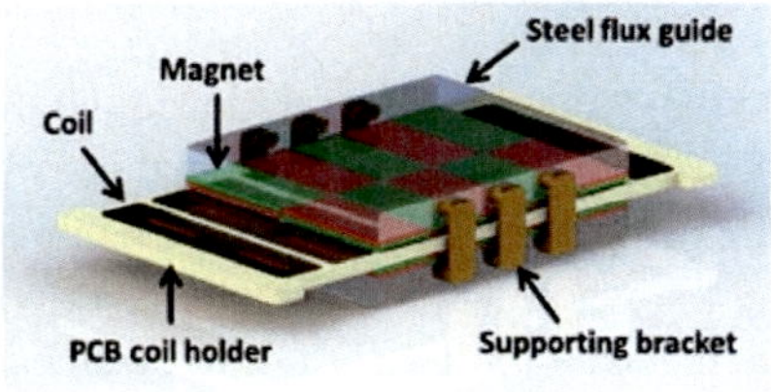

(c) HAC5

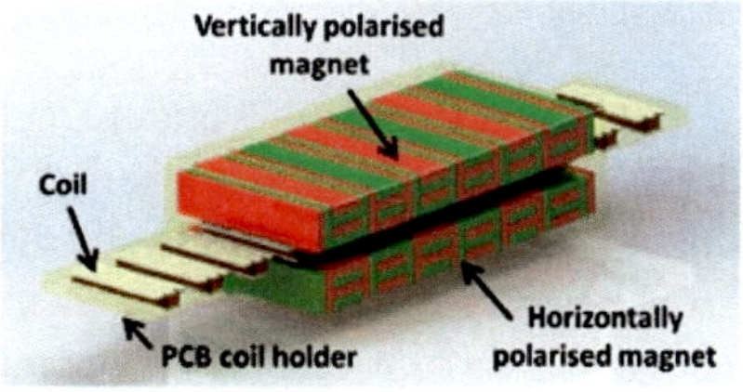

(d) HAC6

Fig. 4.2 CAD Renderings of fabricated harvester architectures that show the magnet and coil placement. Devices are shown in order of their number from HAC1 in pane (**a**), HAC3 in (**b**), HAC5 in (**c**) and HAC6 in pane (**d**). Figures from [9]

component epoxy raisin "Scotch-Weld DP125" from "3M" [4] is applied to each surface and the magnets and spacers are stacked within the gluing aid. A lid is screwed on to keep the magnet stack compressed and the whole setup is placed within an oven where the epoxy resin can cure for 48 h at a steady temperature of 40 °C. In HAC1 the magnets slide within a plastic tube with the coils placed on the outside of the tube as shown in Fig. 4.2a. The coils were wound manually onto the plastic tube, using a prototyping coil winder [5].

In HAC3 the magnets are ring magnets and the magnet stack is built in the same manner as HAC1. However, the coils form a long cylinder and are placed within the ring magnets, that form a tube around the coils (see Fig. 4.2b). The magnet stack slides within the device housing without touching the coil setup.

HAC5 is a structure, which forms a closed magnetic circuit by placing magnets and a back steel sheet on either side of the coil (Fig. 4.2c). The magnets hold on to the back steel sheet due to the attractive force, hence no glue is required. The two parts of the magnetic circuit attract each other strongly and are held apart by small supporting brass brackets on either side. The magnet structure was intended to span the whole available width within the housing. However, for availability reasons a chessboard approach was chosen and the width of the device is now actually spanned by two shorter magnets of 17 mm length. This in turn means that both the number of magnets and the number of coils is doubled to span the device width.

In HAC6 the magnets are placed in a specific manner to form a double Halbach array which is shown in Fig. 4.2d. The unique characteristic of the double Halbach array is that the magnetic field is focused toward the coil space in between the two

Fig. 4.3 HAC6 3D-printed slider that holds the magnetic circuit. It can move freely within the housing and has no contact to the coils on PCB-board within it. Figure from [9]

single Halbach arrays, while the field outside the device is greatly reduced, i.e. to a certain degree it features inherent shielding. The vertically oriented magnets were put together by two magnets of 2 mm height in order to reach the required 4 mm. As was mentioned for HAC5, magnets that span the whole width of the housing were not available (see Table 4.4). The magnets are placed within a 3D-printed slider that holds them at the required orientation. Again no glue is required, however some of the available space is lost to the slider (shown in Fig. 4.3). The height of the magnetic circuit for instance is only 10 mm of the available 12.5 mm.

For both HAC5 and HAC6 coils with rectangular shape were custom fabricated by "Michael Müller Spulenwickeltechnik GmbH & Co.KG". Due to the winding process, the shape of the coils is slightly oval and the total number of windings is not as high as expected using the (idealized) values from copper wire table according to IEC 60317 [6]. This is true for all architectures and has to be taken into account in the model revision in Sect. 4.5.

4.4 Experimental Results

4.4.1 Experimental Setup

In order to measure the voltages generated by the fabricated devices, a mounting aid was fixed to the shoe sole of a running shoe with long screws (shown in Fig. 4.5b). This mounting aid is very robust and one can easily slide the harvesting devices onto it. The measurements are greatly facilitated this way, as the harvesters do not need to be integrated into the shoe sole at this point and modifications can still be made if necessary. The leads of about two meters of length coming off the harvester are fixed to the leg with Velcro straps and inserted into channel one of a NI USB-6211 USB data acquisition card (DAQ). The DAQ card is connected to a laptop and the voltage is recorded using National Instruments' software SignalExpress. A sampling rate of 3000 samples/second was used and a total of 30 s were recorded for each measurement using an ohmic load. Measurements were performed on a treadmill as shown in Fig. 4.4 at motion speeds of 4, 6, 8, and 10 km/h. The load sweeps were recorded at a constant 4 km/h. The voltage graphs shown in this work were all recorded at 4 km/h to allow the comparison of devices and also because it is

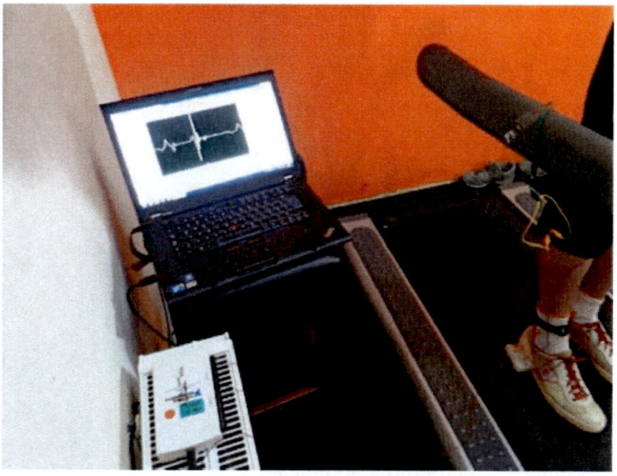

Fig. 4.4 Measurement setup with DAQ card and laptop on the left, treadmill on the right and harvester mounted to the shoe on the bottom right. Figure from [9]

the slowest measured speed and thus the worst case scenario for the swing harvesters in terms of generated energy.

4.4.2 HAC1

This section describes the results for HAC1 in detail (see the fabricated device in Fig. 4.5a). The following sections present the remaining HAC results in an identical manner. All data in this chapter was measured with test person "1". Data for the second test person is given in the discussion in Sect. 4.6. First, the voltage signal and the corresponding power output for a period of 1.5 s is given in Fig. 4.6. This piece of data contains approximately one step and allows a detailed look at the harvester output. The voltage was measured across an ohmic load equal to the total coil resistance of the series-connected coils, 70 Ω in the case of HAC1. The peak voltage is approximately 4 V and the maximum power is 150 mW. The measured average power output at 4 km/h is 4.64 mW.

Figure 4.7 shows the voltage signal and the corresponding power output for HAC1 XL. This is a version of HAC1 that contains larger coils at a height of 10 mm instead of 3.2 mm in the standard version. The two devices are otherwise identical. The standard HAC1 was fabricated as closely as possible to the simulation results from Sect. 3.2. HAC1 XL was fabricated in order to have a second device and a second set of data for the model verification. It is shown in the following that it generates more energy than the standard version of HAC1. This information was useful for revising the system model. The signal was measured across an ohmic load of 220 Ω and the voltage peaks out at approximately 8 V. The power output reaches up to 300 mW depending on the irregular motion of the test person. The average power output generated with this device at 4 km/h is 5.62 mW. The number of peaks corresponds to the number of magnetic flux changes experienced by a coil. In this particular case

(a) HAC1 (b) HAC1 XL

Fig. 4.5 Picture of the fabricated device for the standard version of HAC1 in (**a**) and HAC1 XL mounted to the side of a shoe for the experimental evaluation in (**b**). Notice the larger coils of HAC1 XL. Figures from [9]

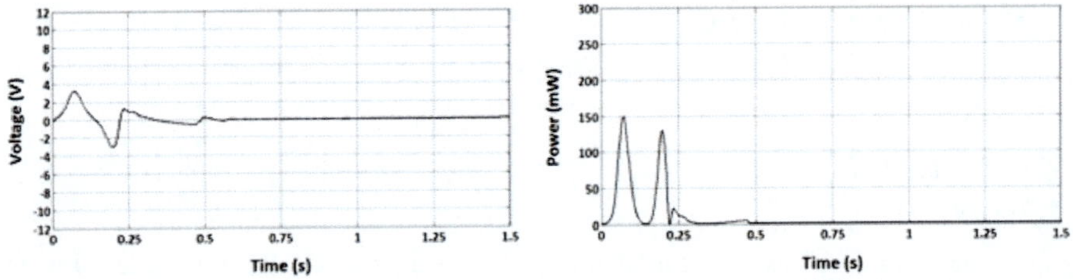

(a) (b)

Fig. 4.6 Voltage output across an ohmic load of 70 Ω in (**a**) and the corresponding power output for HAC1 in (**b**). Measured on a treadmill at a motion speed of 4 km/h. Figures from [9]

there are two changes in the direction of the magnetic flux density in between the three magnets which cause the two major peaks per step in both Figs. 4.6 and 4.7.

In Fig. 4.8 the load sweeps for both versions of HAC1 are given. To obtain this data the treadmill was kept at a constant speed of 4 km/h while single measurements of 30 s length were done for several different load resistances. However, it has to be kept in mind that these curves are only indicative of the actual optimal load as it is highly unlikely that the test person walked every single step absolutely identically.

Figure 4.8a shows the result for the standard version of HAC1 and it can be seen that the optimal ohmic load is in the vicinity of the total coil resistance of 70 Ω. Figure 4.8b on the other hand shows the results for HAC1 XL and in this case the optimal load of approximately 350 Ω appears to be significantly larger than the total coil resistance of 220 Ω. It is assumed, that this deviation from the coil resistance is caused by the difference in electrical damping due to the larger coils. The simulations show, that the maximum electrical damping force of HAC1 XL is half that of HAC1 (at 4 km/h, approx. 0.5 N as opposed to approx. 1 N) due to the increased coil resistance and the consequently lower induced current. Although

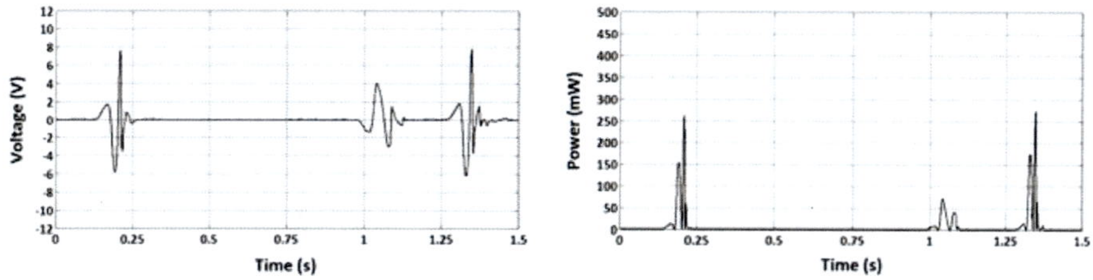

(a) (b)

Fig. 4.7 Voltage output across an ohmic load of 220 Ω in (**a**) and the corresponding power output for HAC1 XL in (**b**). HAC1 XL is a version of HAC1 with larger coils (10 mm coil height instead of 3.2 mm for the standard setup). Measured on a treadmill at a motion speed of 4 km/h. Figures from [9]

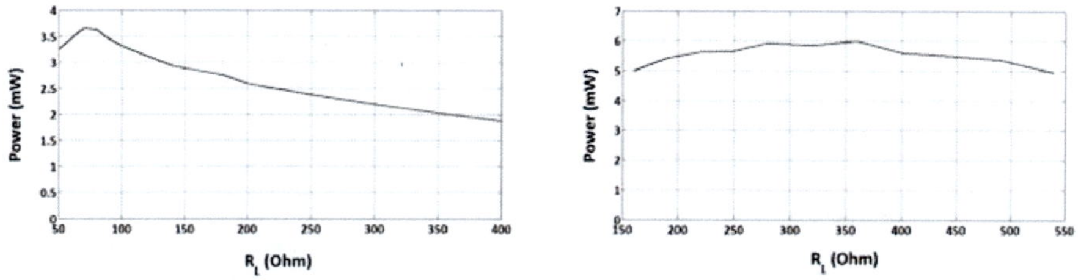

(a) HAC1 (b) HAC1 XL

Fig. 4.8 The load sweep for HAC1 is shown in (**a**) (the total coil resistance of HAC1 is 70 Ω). The load sweep for HAC1 XL is shown in (**b**) (the total coil resistance of HAC1 XL is 220 Ω). Measured on a treadmill at a motion speed of 4 km/h. Figures from [9]

HAC1 XL is not operated in resonance, the optimal load resistance appears to contain a damping-dependent component as is the case with vibration energy harvesters [7]. This was not observed for the standard HAC1 and cannot be measured experimentally. A statistical evaluation across a large number of test persons is required in order to reduce the effects from irregularities of the human gait of the test person for which the graphs are given in this work.

4.4.3 HAC3

The voltage signal recorded for HAC3 at a motion speed of 4 km/h is shown in Fig. 4.9a, while the corresponding power output is given in Fig. 4.9b. Notice that the scale of the graphs was changed significantly as this architecture produces the least amount of energy of all devices presented in this work. The fabricated device is shown in Fig. 4.10a.

The voltage signal was measured across an ohmic load of 33 Ω which is equal to the total coil resistance of the fabricated HAC3 device. The measured voltage peaks

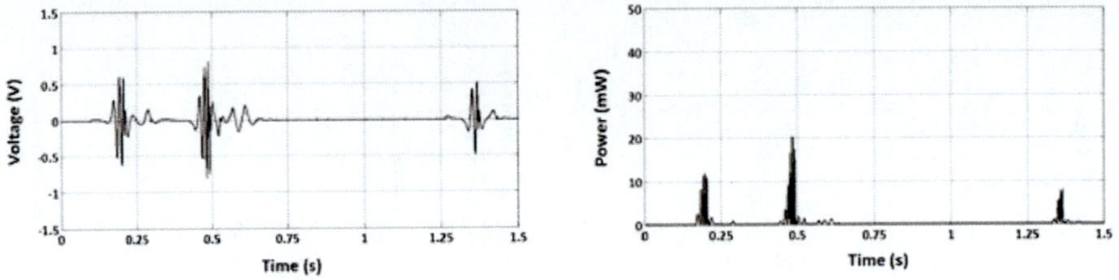

(a) (b)

Fig. 4.9 Voltage output across an ohmic load of 33 Ω in (**a**) and the corresponding power output for HAC3 in (**b**). Measured on a treadmill at a motion speed of 4 km/h. Figures from [9]

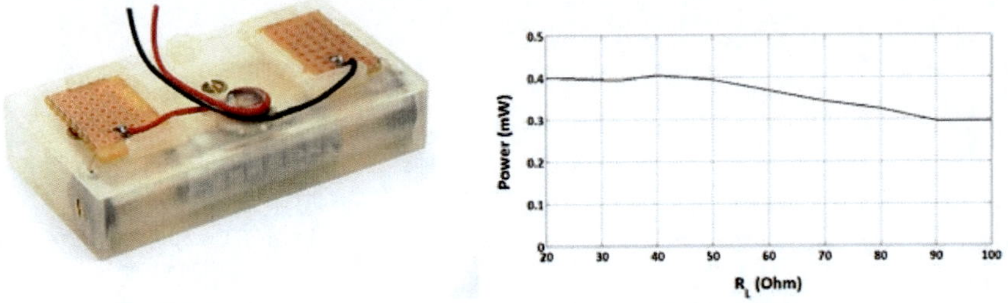

(a) (b)

Fig. 4.10 A picture of the fabricated device for HAC3 is shown in (**a**). The magnet stack can be seen through the 3D-printed housing. The load sweep for HAC3 is shown in (**b**) (the total coil resistance of HAC3 is 33 Ω). Figures from [9]

were lower than one volt and the power output barely reaches peaks of 25 mW. The average power output is 0.39 mW. The load sweep in Fig. 4.10b shows a plateau at loads ranging from 20 to 40 Ω, while the optimal load appears to be 40 Ω which is slightly larger than the coil resistance.

4.4.4 HAC5

HAC5 consists of magnets and back iron parts which form a closed magnetic circuit, i.e. the magnetic field is strong and relatively homogeneous in the coil space between the magnetic parts. A PCB board is used to accommodate the coils which are 1.2 mm high, while the circuit board is 1.6 mm high.

The coils wires are soldered directly onto the PCB and the correct interconnection of the wires is done within the PCB. The coils are held in place at the soldering points and additionally by a 50 μm thick layer of Kapton tape that covers both the top and bottom side of the PCB.

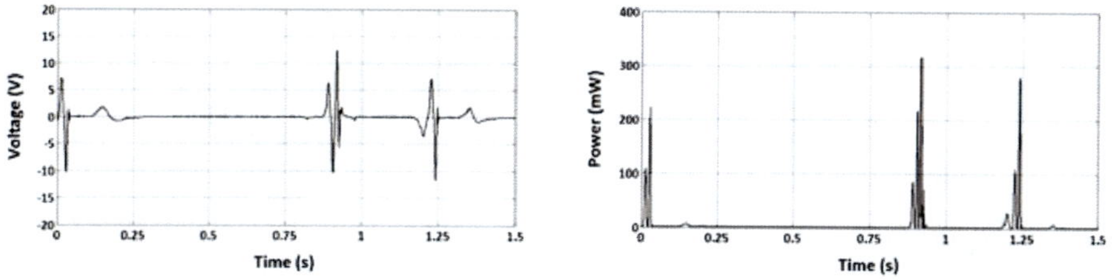

(a) (b)

Fig. 4.11 Voltage output across an ohmic load of 480 Ω in (**a**) and the corresponding power output for HAC5 (**b**). Measured on a treadmill at a motion speed of 4 km/h. Figures from [9]

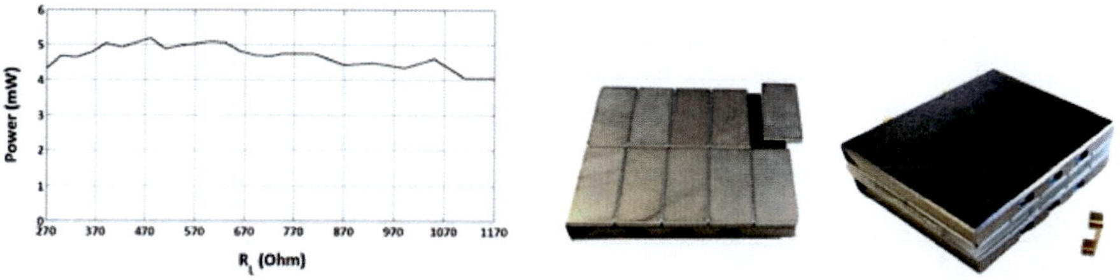

(a) (b)

Fig. 4.12 Load sweep for HAC5 (the total coil resistance of HAC5 is 480 Ω) in (**a**). Measured on a treadmill at a motion speed of 4 km/h. Two exemplary fabrication steps of HAC5 are shown in (**b**). Figures from [9]

The acquired voltage signal is shown in Fig. 4.11. The voltages reach up to 12 V at a motion speed of 4 km/h on the treadmill (Up to 17 V at 6 km/h). The corresponding power peaks reach approximately 450 mW. The average power output of the device is 4.89 mW at 4 km/h.

The magnet stack consists of a chessboard pattern in which on either side of the closed magnetic circuit, 10 magnets are placed in two rows of five magnets (see Fig. 4.12). There are two rows of coils as shown in Fig. 4.13a that correspond to the two rows of magnets.

Due to the five magnet setup there is a total of four magnetic flux changes within the magnet stack, which are experienced by the coils and represented by four peaks in the voltage and power graphs of Fig. 4.11 (as long as the magnets move fast enough to generate a noticeable output).

A load sweep was also performed for the range of ohmic resistances between 270 Ω and 1170 Ω. The resulting graph is given in Fig. 4.12. The largest power output is located approximately at a load equal to the inner coil resistance of 480 Ω while the power output drops towards smaller and larger ohmic loads. The noticeable variations in the graph can be attributed to the human motion patterns which are not ideally repeated from step to step.

An additional version of HAC5 was fabricated using high-tech plastic springs from "Lee Spring Company" (type LL038038) as depicted in of Fig. 4.13b. Whereas the standard version of HAC5 uses a total of 16 coils, here only a total of 12 coils

(a) HAC5 (b) HAC5 F

Fig. 4.13 Picture of fabricated standard version of HAC5 in (**a**). One can notice the circuit board which holds the coils in between the two parts of the magnetic circuit. The rubber end stops were added later. The position of the springs in the modified version of HAC5 is shown in (**b**). The number of coils had to be reduced due to the springs. Figures from [9]

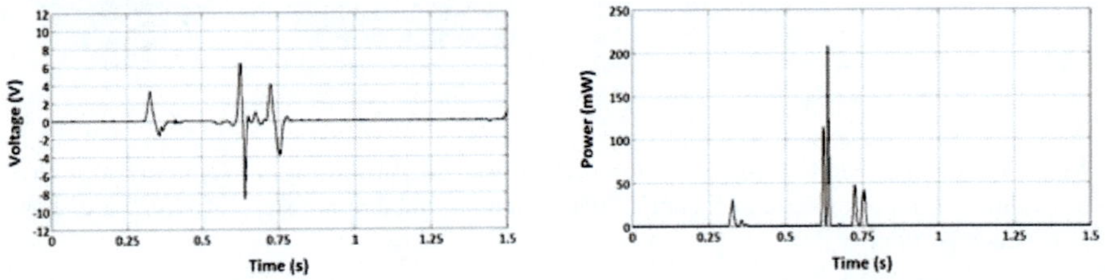

(a) (b)

Fig. 4.14 Voltage output across an ohmic load of 360 Ω in (**a**) and the corresponding power output for HAC5 with springs in (**b**). This device uses only 12 coils instead of the 16 coils in the standard version, thus explaining the lower voltages when compared to Fig. 4.11. Measured on a treadmill at a motion speed of 4 km/h. Figures from [9]

could be accommodated as the available space is reduced due to the springs. In terms of comfort for the user, the addition of springs is a great advantage as one can hardly feel anything moving in the shoe.

The magnetic circuit weighs about 100 g which can certainly be felt, when there are no springs, even though rubber end stops are used. Standard metallic springs could not be placed within the housing, as the magnet stack is attracted to them and the motion is thus severely hindered. In a future setup, the springs might be fitted directly to the magnet stack and move along with it, thus bypassing the effect of attractive magnetic forces. The voltage for HAC5 with springs peaks out at approximately 8 V and the power output can reach values up to 250 mW (see Fig. 4.14). The average power output at 4 km/h is approximately 3 mW.

4.4.5 HAC6

This harvester architecture consists of a double Halbach array. The magnets on either side of the coils are placed in different orientations as shown in Fig. 4.2d in such a way that the intensity of the magnetic field is high within the coil space and low outside the device.

By alternatingly placing horizontally polarized magnets (HPM) next to vertically polarized magnets (VPM), the magnetic flux is guided from one VPM to the next one and from there across the airgap to the VPM on other side of the device [8]. Thus a closed magnetic circuit ist formed without the use of metal flux guides as was the case for HAC5. The magnetic flux in the airgap between two VPM is concentrated and can thus reach values up to 2 T (significantly larger than the remanent flux of permanent magnets).

However, the magnetic flux in the airgap between two HPM, i.e. next to the VPM air gap, ist very low. Therefore it is not immediately clear, if HAC6 has an advantage over HAC5 and experiments are required. The circuit board holding the 12 coils is mounted within the magnetic circuit in the housing. The magnetic stack can move freely without mechanical contact to the coil structure as is the case with HAC5.

The graph of the measured voltage is given in Fig. 4.15 and shows peaks of up to 4 V. The two packs of voltage peaks repeat at the step frequency of 1 Hz. The power output reaches 150 mW in peaks, while the average output is 1.91 mW. The fabricated device is depicted in Fig. 4.16a. Both sides of the magnetic structure are encased in a 3D-printed bracket which can slide freely within the housing.

A load sweep was performed for HAC6 and the resulting graph is shown in Fig. 4.16b. While the maximum power output is located at about 120 Ω which matches the coil resistance of the device, the reader may notice a plateau of nearly constant power output up to a load of 300 Ω. Again it becomes apparent that a larger number of measurements is required in order to produce statistically averaged data, whereas in this particular case the irregularities in the human gait do not produce a clear figure as was the case in Fig. 4.8a.

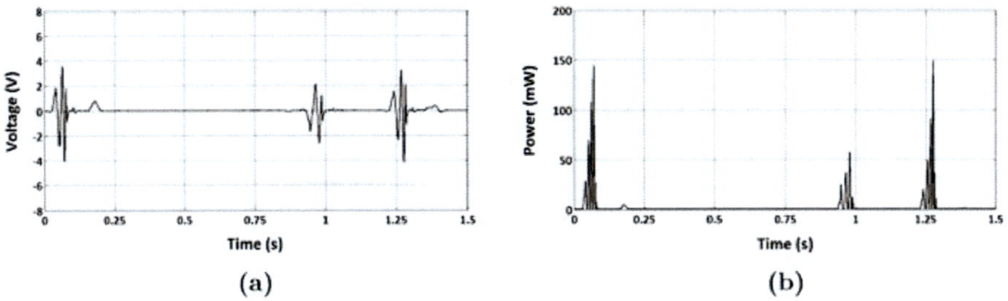

Fig. 4.15 Voltage output across an ohmic load of 118 Ω in (**a**) and the corresponding power output for HAC6 in (**b**). Measured on a treadmill at a motion speed of 4 km/h. Figures from [9]

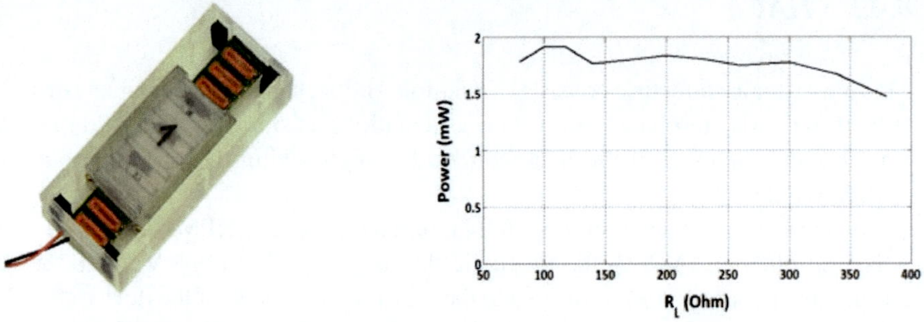

(a) (b)

Fig. 4.16 A picture of the fabricated device for HAC6 is shown on the left. The load sweep for HAC6 is shown on the right (the total coil resistance of HAC6 is 118 Ω). Figures from [9]

4.5 Model Revision and Verification

The system simulations in Matlab/Simulink in the previous chapter were performed using real acceleration data as an excitation input. The available data covers test person "1" and all four motion speeds that were also considered in this work. The comparison between the system models' predictions and the experiment is shown in Table 4.5.

The system model (implemented in Simulink) is identical across all architectures. It is only the simulation of the magnetic field and the preparation of the coupling curve which varies with each architecture. With the help of the experimental data a number of revisions in the evaluation of the magnetic field simulations were performed, while the actual system model remained unchanged. This explains the significantly lower experimental power output when compared to the values in Chap. 3.

Since the experimental devices could not be fabricated to the exact dimensions that were optimized in Chap. 3, e.g. due to the availability of magnets, the simulations were updated with the dimensions of the fabricated devices. The coil parameters in the simulation were updated to produce the resistance of the actual, fabricated coils.

The copper fill factor is a figure that indicates the percentage of a coil's cross-section that is actually filled with copper. This number is far lower in the case of the fabricated coils than was assumed in the simulations, following the information from the IEC coil wire tables [6].

The fabricated coils thus have a smaller number of windings than expected and consequently a smaller resistance. These modification already caused a severe drop in the simulated power outputs as opposed to the previous results presented in Chap. 3.

The reader will notice, that a good match between simulation and experiment has been achieved following these model revisions, particularly at low motion speeds. Table 4.5, shows the full data set for all architectures. All experimental data was obtained using the ohmic loads given in the table (which are equal to the measured total coil resistance and approximately equal to the simulated coil resistance). At the

Table 4.5 Comparison of simulated and measured average power output using an ohmic load equal to the total coil resistance. Because of uncertainties in the coefficient of restitution k_i according to Sect. 2.7, a range of simulated power outputs is given in addition to the values at $k_i = 0.6$ which were used in the optimization process in Chap. 3. The deviations between experiment and simulation are calculated with respect to $k_i = 0.6$. The friction coefficients remain constant as given in Chap. 2. "Sim.", "Exp." and "Dev." indicate simulations, experiments and the deviations respectively. Table from [9]

Device	Type	k_i	4 km/h	6 km/h	8 km/h	10 km/h	Load
HAC1	Sim.	0.5	4.79 mW	7.66 mW	7.58 mW	9.39 mW	70 Ω
		0.6	4.93 mW	8.02 mW	8.04 mW	9.89 mW	
		0.9	5.64 mW	9.66 mW	10.32 mW	12.33 mW	
	Exp.	–	4.64 mW	6.81 mW	6.88 mW	8.57 mW	
	Dev.	0.6	5.88%	15.09%	14.43%	13.35%	
HAC1 XL	Sim.	0.5	5.37 mW	8.81 mW	8.69 mW	11.08 mW	220 Ω
		0.6	5.40 mW	9.02 mW	9.03 mW	11.52 mW	
		0.9	5.72 mW	9.53 mW	10.27 mW	13.28 mW	
	Exp.	–	5.62 mW	9.21 mW	7.52 mW	10.32 mW	
	Dev.	0.6	−4.07%	−2.11%	16.72%	10.42%	
HAC3	Sim.	0.5	0.34 mW	0.57 mW	0.54 mW	0.63 mW	33 Ω
		0.6	0.36 mW	0.70 mW	0.57 mW	0.68 mW	
		0.9	0.52 mW	0.96 mW	1.64 mW	0.93 mW	
	Exp.	–	0.39 mW	0.61 mW	0.57 mW	0.64 mW	
	Dev.	0.6	−8.33%	12.86%	0%	5.88%	
HAC5	Sim.	0.5	4.82 mW	7.83 mW	7.67 mW	8.95 mW	480 Ω
		0.6	5.10 mW	8.79 mW	8.18 mW	9.59 mW	
		0.9	5.96 mW	13.10 mW	10.77 mW	17.48 mW	
	Exp.	–	4.89 mW	8.06 mW	6.79 mW	8.67 mW	
	Dev.	0.6	4.12%	8.30%	16.99%	9.59%	
HAC6	Sim.	0.5	1.84 mW	2.94 mW	3.01 mW	3.51 mW	118 Ω
		0.6	1.92 mW	3.19 mW	3.20 mW	3.76 mW	
		0.9	2.53 mW	4.92 mW	4.06 mW	8.22 mW	
	Exp.	–	1.91 mW	3.26 mW	3.29 mW	4.07 mW	
	Dev.	0.6	0.52%	−2.19%	−2.81%	−8.24%	

motion speed of 4 km/h the simulated and the experimental average power outputs deviate by 5.88%. Generally speaking, the deviations appear to increase with motion speed. Since the motion model produces varying deviations for all architectures, it can be assumed that significant irregularities occur in the gait pattern. It is also assumed that the fabrication process introduces mechanical irregularities and imperfections to the devices. Analogously to the above results, a larger number of measurements is required in the future in order to reduce statistical deviations. An observation to support this theory about fabrication irregularities and gait irregularities is given in Table 4.6, which shows the power outputs recorded for HAC1 using a constant ohmic load of 70 Ω and two different magnet stacks. Herein, only the magnet stack was replaced within the device between measurements, in order to eliminate any

Table 4.6 Ranges of average power output obtained experimentally for HAC1. The coils and the load resistance of 70 Ω remain identical across all measurements, while the magnet stack is changed. The smallest and largest measured values are given for all investigated motion speeds. Table from [9]

	Magnet stack 1		Magnet stack 2	
Measurement	Smallest	Highest	Smallest	Highest
4 km/h	3.59 mW	4.67 mW	3.28 mW	4.60 mW
6 km/h	6.02 mW	6.81 mW	6.09 mW	6.76 mW
8 km/h	4.23 mW	6.10 mW	4.89 mW	6.88 mW
10 km/h	6.07 mW	8.25 mW	6.56 mW	8.57 mW

influences by the coil. The table contains the lowest and highest measured power output for the two magnet stacks, all investigated motion speeds and test person “1”.

The first noticeable fact is that the generated power can vary greatly from one measurement to another as indicated by the variations in power output in Table 4.6, even without changing any parts of the device. This confirms the irregularities in the human gait and shows how a variation in step length and step frequency can have an effect on the harvested energy. The second observation is that there can be great variations in the fabrication of these prototype devices. Although all magnet stacks for HAC1 were fabricated following the procedure in Sect. 4.3, there are differences in power output when the data for the two magnets stacks shown in Table 4.6 are compared. It is particularly the friction of the magnet stack within the plastic tube that appears to be affected by small misalignments of the magnets during the gluing process. In these cases the three magnets and two steel spacers are not ideally concentric, meaning that one of these parts might scratch along the plastic tube and consequently a reduced power output results due to the increased friction.

In the case of HAC1 XL and a motion speed of 4 km/h the deviation between the simulated and the measured power output is −4.07% as shown in Table 4.5. The maximum deviation of 16.72% is found at the motion speed of 8 km/h. In the case of HAC3, the difference between the simulated and the measured power output amounts to −8.33% at 4 km/h with the maximum difference of 12.86% to be found at 6 km/h. The simulation of HAC5 matches the measured value quite well with a deviation of only 4.12% for a motion speed of 4 km/h, with the largest deviation of 16.99% being found for 8 km/h. Lastly, HAC6 also shows a very close match between simulation and experiment, with a deviation of only 0.52% at 4 km/h and a maximum deviation of −8.24% at 10 km/h.

The system model presented in this work contains a number of simplifications, e.g. in the friction model and the modeling of the hardstops, when the magnets reach the end of their motion range. Therefore, this is a satisfactory match between simulations and experiments. One common cause for deviations shall still be mentioned: the representation of the coils in the computer model. Based on the coils dimensions, the program calculates the number of windings and from that the coil resistance. In this part rounding can cause a change in the number of windings and resistance, leading to deviations.

4.6 Discussion

In the previous sections, a number of harvester architectures were investigated and their respective performance was determined experimentally using a treadmill. The resulting average power output for all architectures, for both test persons and for the four analyzed motion speeds is summarized in Table 4.7. In each case, the voltages leading to these power outputs were measured across an ohmic load equal to the respective coil resistance. A dash indicates a missing measurement where a device was broken. The harvester of HAC1 XL performs well and generates an average power output of up to 10 mW.

The power densities for the motion speeds of 4 km/h and 6 km/h are listed in Table 4.8, which contains the average power output at the given speed. In the case of the two cylindrical architectures, two full magnet and coil stacks can be placed within the housing in order to make good use of the width available in the shoe sole. However, experiments with two stacks within one housing have not been performed to this point. These values are theoretical, assuming that the power output of a single stack can be doubled that way. The tables also contain the total volume including the housing and the corresponding power density, as well as the active volume, which is the volume filled by the parts relevant to the transduction process.

In accordance with the data from Table 4.7, the cylindrical architectures of HAC1 show the largest power densities with up to 627 μW/cm^3 at 4 km/h and up to 1027.9 μW/cm^3 at 6 km/h with respect to the active volume. HAC5 makes full use of the width of the housing, hence the large active volume listed in Table 4.8. Although the

Table 4.7 Overview of all performed measurements. "1" and "2" denote the test persons. Measurements were performed at four motion speeds. A dash indicates a missing measurement where a device was broken. Table from [9]

	4 km/h		6 km/h		8 km/h		10 km/h	
	"1"	"2"	"1"	"2"	"1"	"2"	"1"	"2"
HAC1	4.64 mW	2.62 mW	6.81 mW	3.94 mW	6.88 mW	4.62 mW	8.57 mW	5.24 mW
HAC1 XL	5.62 mW	–	9.21 mW	–	7.52 mW	–	10.32 mW	–
HAC3	0.39 mW	0.29 mW	0.61 mW	0.44 mW	0.57 mW	0.50 mW	0.64 mW	0.52 mW
HAC5	4.89 mW	4.47 mW	8.06 mW	6.91 mW	6.79 mW	7.14 mW	8.67 mW	8.56 mW
HAC5 w. springs	3.02 mW	3.07 mW	5.32 mW	4.08 mW	4.83 mW	4.60 mW	6.54 mW	5.52 mW
HAC6	1.91 mW	1.95 mW	3.26 mW	2.96 mW	3.29 mW	3.47 mW	4.07 mW	3.44 mW

Table 4.8 Power output and power densities of all architectures for motion speeds of 4 km/h and 6 km/h for test person "1". In the case of cylindrical architectures HAC1 and HAC3, two complete magnet stacks fit within the housing, although power is lost to a certain degree due to the attractive force between magnet stacks. Table from [9]

	Speed	HAC1	HAC1 XL	HAC3	HAC5 w/o springs	HAC5 w. springs	HAC6
Power output (mW)	4 km/h	4.64	5.62	0.39	4.89	3.02	1.91
Power output (mW) (two stacks per housing)		9.28	11.24	0.78	–	–	–
Total volume (cm^3) (one stack/two stacks)		20.01 / 50.33	20.01 / 50.33	20.01 / 50.33	50.33 /-	50.33 /-	40.02 /-
Power density ($\mu W/cm^3$) (one stack/two stacks)		231.9 / 184.4	280.9 / 223.3	19.5 / 15.5	97.2 /-	60 /-	47.7 /-
Active volume (cm^3) (one stack)		8.96	8.96	8.96	34.22	34.22	22.81
Power density ($\mu W/cm^3$)		517.9	627.2	43.5	142.9	88.3	83.7
Power output (mW)	6 km/h	6.81	9.21	0.61	8.06	5.32	3.26
Power output (mW) (two stacks per housing)		13.62	18.42	1.22	–	–	–
Total volume (cm^3) (one stack/two stacks)		20.01 / 50.33	20.01 / 50.33	20.01 / 50.33	50.33 /-	50.33 /-	40.02 /-
Power density ($\mu W/cm^3$) (one stack/two stacks)		340.3 / 270.6	460.3 / 366	30.5 / 24.2	160.1 /-	105.7 /-	81.5 /-
Active volume [cm^3] (one stack)		8.96	8.96	8.96	34.22	34.22	22.81
Power density [$\mu W/cm^3$]		760.0	1027.9	68.1	235.5	155.5	142.9

average power output is quite high, the power density of HAC5 is lower than is the case for either version of HAC1. In the case of HAC6, the housing is a bit narrower, as it was adapted to the magnets, which were not available in a length that could span the whole width of the standard housing. HAC 6 performs worse than the other cuboid devices and overall it is second to last in terms of both absolute power output and power density. HAC3 shows the worst performance of all investigated devices. One possible reason is that the coils are very small as they need to fit within the magnet rings used in this architecture and therefore only a small number of windings can contribute to the power output.

The investigated devices show different behaviors as far as the test persons are concerned. It is obvious that the power output of a device will vary with the test person and the corresponding gait pattern. The variations however, show a certain pattern. For those devices, where the magnet mass is comparatively small, e.g. HAC1 and HAC3, the power output for test person "1" is larger than for "2" throughout all measurements. In some cases, the power output is almost twice that of "2". In the case of the architectures with heavier magnet stacks, e.g. HAC5 and HAC6, the difference in power output between the two test persons is relatively small. For the measurements performed at 8 km/h in fact, the order is reversed and test person "2" generates a larger power output than "1". This observation indicates the importance of the inertial mass in this kind of system. With a larger inertial mass, the difference in the gait pattern appears to lose importance. The kinetic energy is proportional to the moving mass and it follows that the damping of the motion, both mechanical and electrical, has a reduced relative effect on magnet motion as the mass increases.

When considering user comfort, the magnets of HAC1 with their relatively low mass of approximately 18.4 g per magnet stack, subjectively do not create a disturbance in the gait pattern. In fact they can hardly be felt. The heavy magnetic circuit of HAC5 however can clearly be felt, even with rubber end stops. The addition of plastic springs solved this problem at the expense of a significantly reduced power output (see Table 4.7), as the number of implemented coils had to be reduced to accommodate the springs. In addition, the longevity of these springs is highly questionable. They have to withstand strong impacts for an extended period of time. During the treadmill experiments, the springs broke after a few measurement sessions, effectively amounting to a life-time of only a few dozen minutes during these tests.

4.7 Outlook - Additional Experiments

A small series of additional experiments was performed using HAC1 XL as a basis, as it proved to be the most suitable device so far. These additional experiments aim to show that mechanical improvements to the harvesting devices can be achieved through several different approaches. In Sect. 4.5 experimental data was given, which showed how small changes in the gluing process of the magnets can cause variations in the power output.

The optimization of the geometrical parameters as performed in Chap. 3 is the obvious path towards a maximum power output. The third option is to improve the number of coil windings within a given volume by modifying the winding process. The fourth option is to parallelize, where possible. The latter two options are briefly outlined in the following, while several further ways of improving this type of device may exist.

For the modified test device HAC1 XL-N, the coils were ordered in the same dimensions as before, but since they were fabricated industrially, the resistance is not 220 Ω but 280 Ω instead. This indicates, that the number of windings is significantly larger than was the case for the prototype coils of HAC1 XL. A larger number of coil windings implies a better coupling and an improved power output can be expected.

As expected, the power output of the device using improved coils is larger. At 4 km/h for test person "1", an average power output of 6.01 mW is measured using an ohmic load of 280 Ω (see Fig. 4.17 for the voltage and power output graphs). At 10 km/h this value increases to 11.48 mW.

The increase in power output over the standard version of HAC1 XL, considering all four measured motion speeds, lies in the range between 6.49% and 17.09%. This was achieved solely by means of an improved winding process.

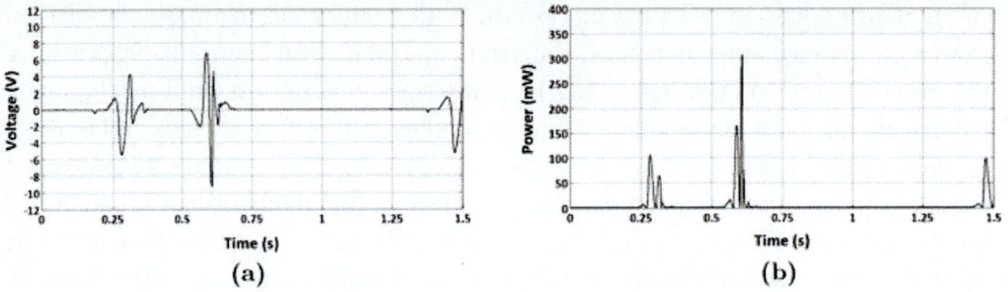

Fig. 4.17 Voltage output across an ohmic load of 280 Ω in (**a**) and the corresponding power output for HAC1 XL-N in (**b**). HAC1 XL-N is a version of HAC1 XL with industrially fabricated coils that have a larger number of windings in the same volume and a total resistance of 280 Ω instead of 220 Ω. Measured on a treadmill at a motion speed of 4 km/h. Figures from [9]

Parallelization was mentioned as the fourth option to improve the power output. This is only possible in the case of cylindrical architectures where two magnet-coil-setups can be placed next to each other within the same housing. The housing dimensions remain identical to all previous considerations of this work. A harvesting device with two stacks of HAC1 XL-N, i.e. with the newly fabricated coils, is evaluated experimentally. The diameter of one magnet coil setup is 12.5 mm which translates into a total maximum of three stacks next to each other given the 37.5 mm of width available within the housing. The space for the middle stack is left empty in order to leave some distance between the two outer stacks. Although there is a distance of more than a centimeter between magnets stacks, during fabrication a noticeable attractive force between the magnet stacks was observed. This attractive force pulls the stacks towards one another leading to a significant increase in friction.

Figure 4.18 shows the voltage graphs obtained for the four motion speeds that are investigated in this work and test person "1". It can immediately be noticed that during the smooth motions of running at 8 km/h and 10 km/h, hardly any voltage is generated. This implies that the external accelerations due to the gait pattern were not large enough to counter the friction due to the attractive force between magnet stacks. This observation is supported by the average power output figures given in Table 4.9. At 4 km/h the generated voltage and thus the power output is still larger than for the running speeds of 8 km/h and 10 km/h. The largest voltage peaks occur at a speed of 6 km/h, where subjectively the leg motion is very abrupt and thus the magnet stacks can overcome the friction force.

When comparing the single-stack device to the two-stack device, it can be noticed that even though twice as many magnets and coils are used, the power output does not double. In fact, only at a motion speed of 6 km/h does the two-stack device show any advantage over a single-stack device, i.e. the sum of the power outputs of the two parallel stacks is larger than for the single stack. At the motion speeds of 8 km/h and 10 km/h in particular, hardly any energy can be generated and a single-stack device can produce more than 5 times as much power (see Table 4.9).

At 6 km/h the leg motion is rougher and more abrupt than when a person is running, thus the friction force is overcome and the power output increases. However, from

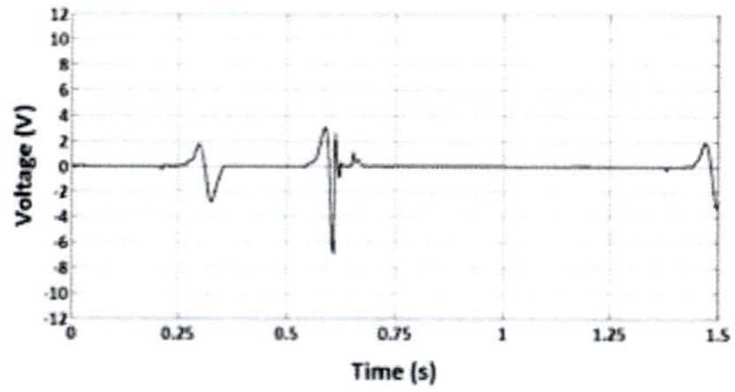

(a) 4 km/h

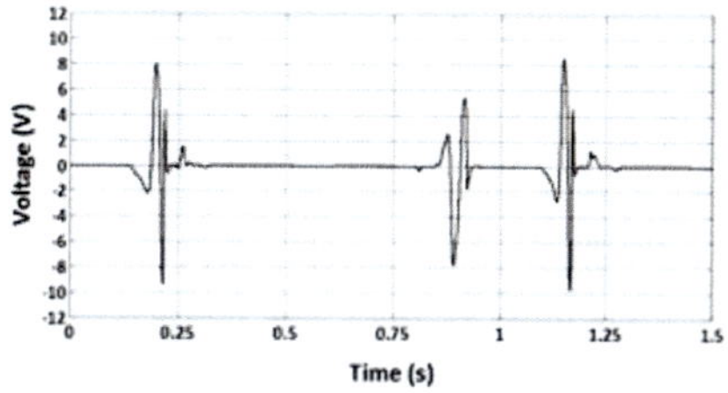

(b) 6 km/h

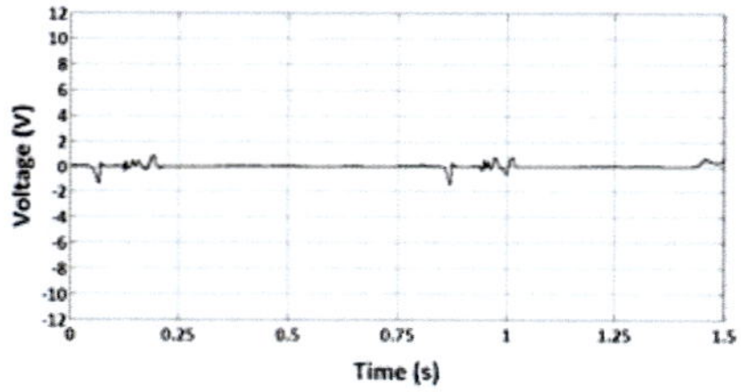

(c) 8 km/h

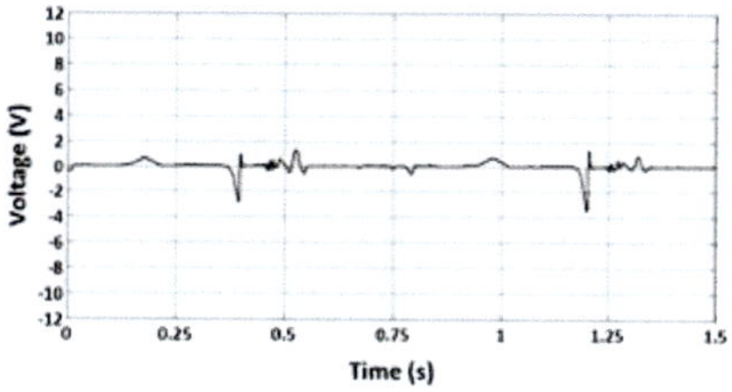

(d) 10 km/h

Fig. 4.18 Voltage data obtained experimentally for HAC1 XL-N with two parallel magnet stacks within one housing for test person "1". The data is shown for only one of the two stacks. The coils and the load resistance of 284 Ω per stack remain identical across all measurements. Figures from [9]

Table 4.9 Average power output measured experimentally for HAC1 XL-N with one or two parallel magnet stacks within one housing. The data is given for four motion speeds and two runners. "1/2" and "2/2" indicates the power output for the first and second of the two parallel stacks. Table from [9]

	Person "1"			Person "2"		
	Single stack	Stack 1/2	Stack 2/2	Single stack	Stack 1/2	Stack 2/2
4 km/h	6.01 mW	2.02 mW	1.95 mW	4.23 mW	1.20 mW	1.16 mW
6 km/h	10.72 mW	9.01 mW	9.02 mW	6.71 mW	3.56 mW	3.57 mW
8 km/h	9.07 mW	0.15 mW	0.17 mW	5.91 mW	0.30 mW	0.27 mW
10 km/h	11.48 mW	0.95 mW	1.09 mW	10.11 mW	0.96 mW	0.93 mW

the data in Table 4.9 it can be observed, that this also greatly depends on the runner. For test person "2" the power output of the two-stack device approximately matches the single-stack device at 6 km/h, while for test person "1", about twice the power is generated compared to the single-stack case.

References

1. K. Ylli, D. Hoffmann, B. Folkmer, Y. Manoli, Design, fabrication and characterization of an inductive human motion energy harvester for application in shoes. J. Phys.: Conf. Ser. **476**, 012012 (2013)
2. K. Ylli, D. Hoffmann, P. Becker, A. Willmann, B. Folkmer, Y. Manoli, Human motion energy harvesting for aal applications. J. Phys.: Conf. Ser. **557**, 012024 (2014)
3. K. Ylli, D. Hoffmann, A. Willmann, P. Becker, B. Folkmer, Y. Manoli, Energy harvesting from human motion: exploiting swing and shock excitations. Smart Mater. Struct. **24**(2), 025029 (2015)
4. 3M, Scotch-weld dp125 (2017), http://kleben.3mdeutschland.de/produktkatalog/klebstoffe/3m-scotch-weld-2-komponenten-konstruktionsklebstoffe-fuer-das-epx-system/epoxidharz-klebstoffe/3m-scotch-weld-2k-konstruktionsklebstoff-fuer-das-epx-system-dp-125.html. Accessed 03 April 2017
5. CNCDesign, Cncdesign 200mm coil winder (2017), http://www.ukcnc.net/index.php/products/200mm-coil-winder/ Accessed 03 April 2017
6. Elektrisola, Iec 60317 standard - technical data for self bonding copper wire (2017), http://www.elektrisola.com/de/backlackdraht/technische-daten-nach-durchmessern/iec-60317-ueberblick.html. Accessed 03 April 2017
7. D. Spreemann, Y. Manoli, Electromagnetic vibration energy harvesting devices: Architectures, design, modeling and optimization, vol. 35 of Springer series in advanced microelectronics. Springer, Dordrecht, New York (2012)
8. D. Zhu, S. Beeby, J. Tudor, N. Harris, Vibration energy harvesting using the Halbach array. Smart Mater. Struct. **21**(7), 075020 (2012)
9. K. Ylli, Energy harvesting from the swing motion of the human leg (2019)

Chapter 5
Second Optimization Run

Building on the results of the previous chapters, the system model was revised using experimentally obtained data from the four harvester architectures under investigation (see Chap. 4). It can be assumed, that the revised model will produce optimal geometrical parameters, which differ from the previous results presented in Chap. 3. In order to find the optimal geometry for the given available volume, a second optimization run is performed using the revised system model.

The parameter boundaries for each architecture remain identical to the values given in Chap. 3 and the optimization constraints given in Eqs. 3.1 to 3.15 are still valid as well. The following sections briefly present the results of the respective optimization runs.

5.1 HAC1

The results for the new optimization of HAC1 are listed in Table 5.1 as the dataset "O2". "O1" indicates the previous optimization results and "E" the design and experimental data from the fabricated device. This table is given for all architectures in an equivalent form and all power output data corresponds to a motion speed of 4 km/h, both in simulations and experiments on a treadmill.

It was mentioned in Sect. 4.3 that commercially available magnets were chosen to fabricate the investigated harvesters. They were chosen to closely match the desired dimensions for "O1". Nevertheless, due to change in magnet dimensions, new simulations were performed to optimize the remaining parameters. This for example explains the change in the coil length between "O1" and "E". The fabrication of the other HACs was handled accordingly.

Analogously to the previous simulations, the magnet radius in "O2" is chosen as large as possible within the given boundaries and the magnet has to be a full cylinder

K. Ylli and Y. Manoli, *Energy Harvesting for Wearable Sensor Systems*,
Springer Series in Advanced Microelectronics 62,
https://doi.org/10.1007/978-981-33-4448-8_5

Table 5.1 Geometrical Parameters and average power output of the best design found for HAC1. The data is given for optimization run "O1" and the corresponding experiment "E" as well as the second optimization run "O2". All lengths are given in mm. Table from [1]

	R_{mag}	r_{mag}	h_{mag}	l_{coil}	d_{coil}	d_{Spacer}	n_{mag}	n_{coils}	P_{RMS} (mW)
O1	4.5	0	15.55	15.95	0.5	2.2	2	4	8.28
E	4.5	0	12.0	3.2	1.0	1.4	3	4	4.64
O2	4.45	0	7.0	6.15	1.05	1.5	5	7	7.39

and not a ring. The magnet also is not as long as before with a simulated height of 7 mm. This in turn frees up space and more magnets can be used to form one magnet stack.

Consequently, the number of coils increases with the number of magnets. The coils in "O2" are shorter than in the previous simulations and approximately as long as the magnets. With the coils being about as long as the magnets, the space within the housing is used very effectively and the electrical damping of the magnet motion due to relatively large coils does not negatively impact the power output.

The power output of "O2" has dropped over "O1" due to the model revision. The expected power output of "O2", however, is larger than was the case with the experiments of HAC1, indicating that the power output for this architecture could be improved further.

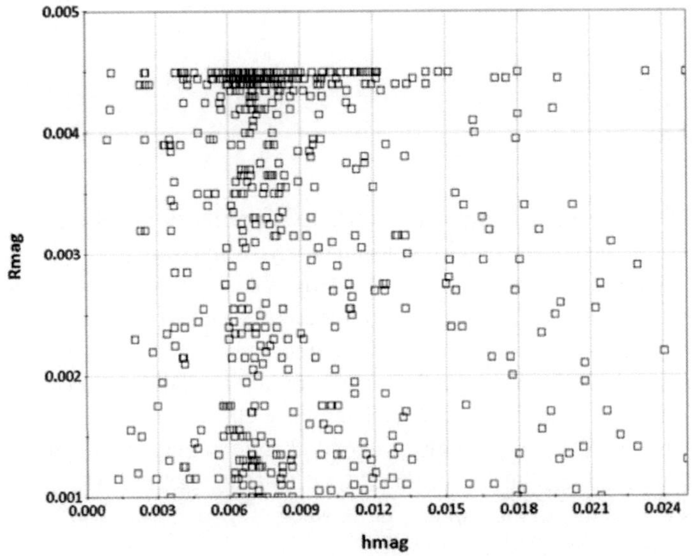

Fig. 5.1 Scatter graph showing the distribution of the designs found for HAC1 during the optimization process along the height and radius of the magnet. Lengths scale in meters. Figure from [1]

The scatter graph in Fig. 5.1 shows the distribution of the designs, i.e. sets of geometrical parameters, found during the second optimization run. Designs on the margins of the graph, e.g. towards a large h_{mag} generate small power outputs and are quickly discarded by the algorithm. The algorithm generated a total of 855 feasible designs within the boundaries, most of which gather around the magnet heights of 6 mm to 9 mm. A large R_{mag} is advantageous to the power output, explaining the large number of designs at the upper margin of the parameter. Scatter graphs showing the relation between designs found by the optimization algorithm and the coil parameters can be found in Sect. A.2 for all harvester architectures.

5.2 HAC3

The optimization process is repeated for HAC3 with the corresponding boundaries, leading to the results in Table 5.2. The power outputs in Table 5.2 show a large difference between the experimental power output and the predicted value from the first optimization run "O1". A first reason is, that the fabricated device does not directly match the designs found by the optimization. This already causes a difference in power output. After the revision and verification of the system model, the simulated power outputs closely matched the experiments. Using the revised model, "O2" produces an expected power output that is 3.5 times larger than the recent experimental value.

The new set of geometrical parameters "O2" is similar to the previous set "O1" in terms of magnet dimensions and the number of magnets and coils used. The coils are significantly larger however, as indicated by the coil length l_{coil}. The scatter graph in Fig. 5.2 shows the distribution of the designs found by the optimization algorithm. The largest number of designs is gathered around magnet heights of 3 mm to 4 mm. This shows that larger magnets do not necessarily generate a larger power output. By using several magnets in opposing polarization in a row, an increased number of flux changes occur during each motion phase. The graph shows, that there is an optimum balance between the magnet size and the number of magnets used.

Table 5.2 Geometrical Parameters and average power output of the best design found for HAC3. The data is given for optimization run "O1" and the corresponding experiment "E" as well as the second optimization run "O2". All lengths are given in mm. Table from [1]

	d_{mag}	d_{iron}	h_{mag}	l_{coil}	d_{coil}	d_{Spacer}	n_{mag}	n_{coils}	P_{RMS} (mW)
O1	2.9	0	3.4	1.65	2.05	1.0	9	12	7.05
E	2.25	0	4.0	1.0	2.0	0.5	8	9	0.39
O2	2.55	0	3.65	4.05	2.4	1.0	9	10	1.43

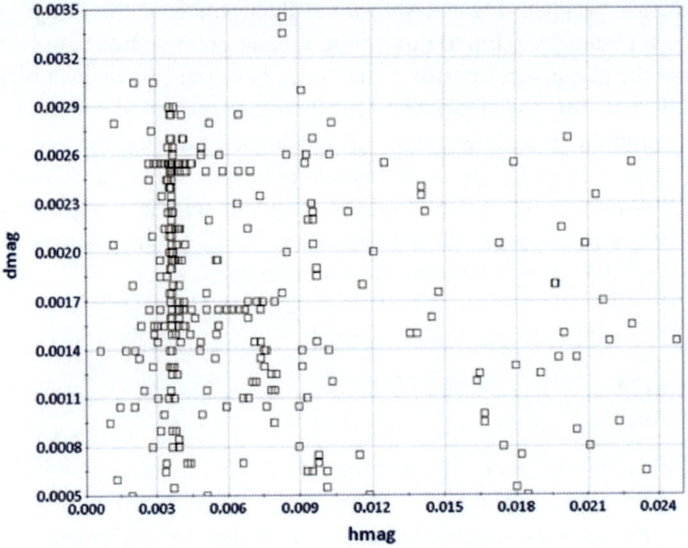

Fig. 5.2 Scatter graph showing the distribution of the designs found for HAC3 during the optimization process along the height and wall thickness of the ring magnet. Lengths scale in meters. Figure from [1]

5.3 HAC5

The deviation in power output between the first optimization run and the experimental characterization for HAC5 is significant as can be seen in Table 5.3. Partially, this can be attributed to the commercial availability of magnets. No magnets were found that could span the whole width of the device housing. Instead magnets are used in the experiment which span half the width of the housing. The magnets are arranged in a chessboard pattern as shown in Chap. 4. This doubles the number of magnets per device and the coils are chosen to match the magnet size.

Table 5.3 Geometrical Parameters and average power output of the best design found for HAC5. The data is given for optimization run "O1" and the corresponding experiment "E" as well as the second optimization run "O2". All lengths are given in mm. Table from [1]

	b_{mag}	d_{iron}	h_{mag}	d_{coil}	h_{coil}	n_{mag}	n_{coils}	P_{RMS} (mW)
O1	1.25	3.8	16.6	1.35	9.8	2	5	42.66
E	1.0	4.0	8.0	1.2	6.5	20	16	4.89
O2	2.65	1.8	10.35	2.55	10.0	4	5	22.00

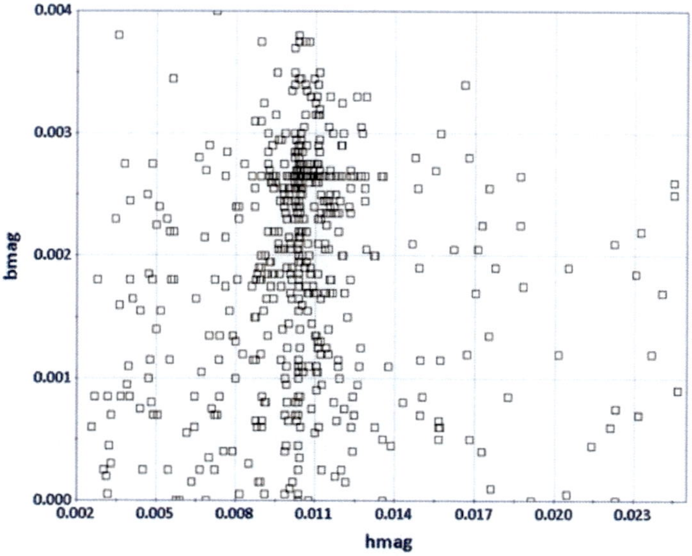

Fig. 5.3 Scatter graph showing the distribution of the designs found for HAC5 during the optimization process along the width (h_{mag}) and thickness (b_{mag}) of the ring magnet. Lengths scale in meters. Figure from [1]

The second optimization run "O2" uses the revised system model. While the expected power output is lower than for "O1", it is noticeably larger than for the experimental device of HAC5.

The ratio of magnet thickness b_{mag} to back steel thickness d_{iron} shows the largest change. In the prior optimization and experiment, the magnet was very flat at a thickness of approximately 1 mm, while the steel back plate that is used as a magnetic flux guide was comparatively thick at 4 mm. This ratio has changed towards a larger magnet, which in "O2" is 2.65 mm thick as opposed to the steel back plate of 1.8 mm thickness.

The scatter graph in Fig. 5.3 shows the designs found by the optimization algorithm against the magnet width h_{mag} and thickness b_{mag}. In terms of the thickness, most designs appear to cluster between 2 mm and 3 mm. In terms of the width, most designs can be found in a narrow band around $h_{mag} \approx 10$ mm.

5.4 HAC6

A comparison between the first optimization run "O1" and the experimental implementation "E" shows a large difference in the size of the magnets and the number of magnets used. The large number of magnets used in the experiment can be explained

Table 5.4 Geometrical Parameters and average power output of the best design found for HAC6. The data is given for optimization run "O1" and the corresponding experiment "E" as well as the second optimization run "O2". All lengths are given in mm. Table from [1]

	b_{mag}	h_{mag}	$h_{mag,2}$	d_{coil}	h_{coil}	n_{mag}	n_{coils}	P_{RMS} (mW)
O1	5.0	14.8	13	1.35	9.1	6	1	23.2
E	4.0	2.0	4.0	1.2	3.8	38	12	1.91
O2	5.0	5.05	5.45	1.25	9.25	14	7	22.9

by the availability of magnets. The magnet width $h_{mag,2}$ is built by placing two magnets of width h_{mag} on top of each other to achieve the required thickness of b_{mag}.

The setup as a double Halbach array requires an additional set of magnets, by building two separate Halbach arrays that are placed within a 3D-printed slider and produce a large magnetic flux in the coils between the single arrays (Table 5.4).

As was the case with HAC5, the revisions of the model for HAC6 mainly affected the magnetic flux calculation and the computation of the coupling coefficient. Hence a large difference in magnet dimensions is found when comparing O2 and O1. The second optimization run "O2" shows that magnet thickness and width are nearly identical. The number of required magnets is smaller than in "E" and the number of coils as well. Additionally the magnets used in the experiment are far shorter than

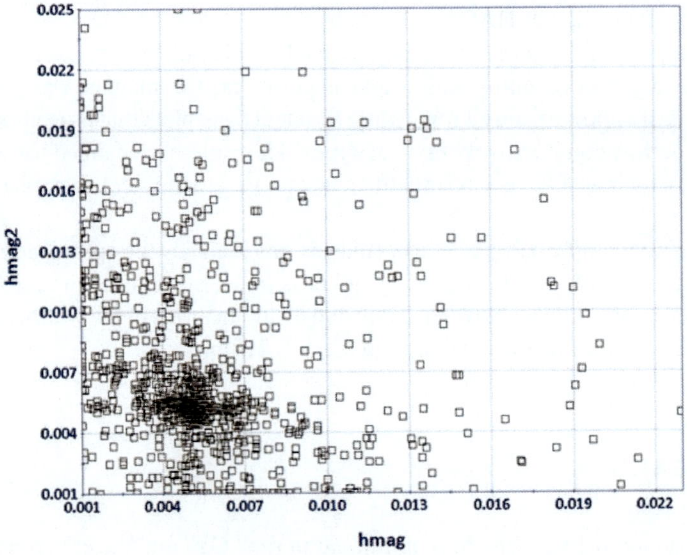

Fig. 5.4 Scatter graph showing the distribution of the designs found for HAC6 during the optimization process along the widths (h_{mag}) and (h_{mag2}) of the two magnet types used in the Halbach array. Lengths scale in meters. Figure from [1]

the full span of 37.5 mm that is used in the optimizations. This deviation, along with the unrevised system model explain the difference between "O1" and "E". The coils are of nearly the same size in both "O1" and "O2" and the coil dimension in the optimizations is approximately 2.5 times the coil dimension in the experiment "E".

Figure 5.4 shows the relation between the width h_{mag} of the two magnet types used and the designs found by the optimization algorithm. It is clear that designs towards the boundary of either magnet are quickly neglected by the algorithm. Most designs gather around the widths of approximately 5 mm. As seen in "O2", the magnets that generate the largest power output in the optimization are of nearly identical size.

5.5 Discussion

The second optimization of the geometrical parameters of the human swing motion harvesting architectures was presented in Chap. 5. Table 5.5 summarizes the optimization results.

Compared to the values in Sect. 3.1, the expected power output has decreased across all architectures after the model revision. The new simulation results however, are still higher than the power outputs from the experimentally analyzed devices. This indicates that there is still space for an improvement of the fabricated devices.

Table 5.5 Summary of the optimized designs for architectures HAC1 to HAC6. In the case of cylindrical HACs, the "total" volume refers to the volume defined in Sect. 1.3. Table from [1]

Architecture	HAC1	HAC3	HAC5	HAC6
Type	Cylindrical	Cylindrical	Rectangular	Rectangular
P_{RMS} (mW) (per magnet stack)	7.39	1.43	22.0	22.9
$P_{RMS,total}$ (mW) (two stacks)	14.78	2.86	–	–
Active volume (cm^3) (one stack)	8.96	8.96	34.22	34.22
Total volume (cm^3) (incl. housing)	50.33	50.33	50.33	50.33
Power density (µW/cm^3) (active volume)	824.8	159.6	642.9	669.2
Power density (µW/cm^3) (total volume)	293.7	56.8	437.1	455.0

The rectangular architectures in particular still have great potential for up to a 5-fold increase in power output for HAC5 and a 12-fold increase for HAC6. The power generation of the fabricated HAC1 is already near the newly calculated optimum with a potential for 1.6-fold increase in power output. For HAC3 there is the potential for a 3.7-fold increase in power output.

The rectangular architectures are expected to generate a larger power output than the cylindrical architectures. The predicted average power output of both HAC5 and HAC6 is above 20 mW.

Reference

1. K. Ylli, Energy harvesting from the swing motion of the human leg (2019)

Chapter 6
Second Generation HAC Experimental Results

In this chapter, the results of the final set of experiments are presented. These results are based on a new and final generation of fabricated harvesters. These devices are based only on architectures HAC5 and HAC6 for two main reasons: the first reason is that the revised system model predicted a far higher power output for these two architectures than for HAC1 and HAC3 (see Chap. 5). The second reason is, as was explained in Chap. 4, that these architectures do not contain magnets in opposing orientation and are thus comparatively easy to assemble due to the lack of opposing magnetic forces.

Unless explicitly mentioned, this chapter presents the measurement results during walking of test person "1" on a treadmill.

6.1 Fabrication HAC5

The results of the optimization for the final design of HAC5 were shown in the previous chapter. Based on those calculated dimensions the CAD design is performed, while staying in constant contact with the workshop in order to receive feedback about whether certain shapes and dimensions can be fabricated at all. This is an iterative process which proved, that the ideal dimensions of the optimization run could not be fabricated exactly as required. Instead the CAD design was modified over a total of six iterations to produce a result that could be fabricated and generate a large power output. Table 6.1 shows two examplary iterations as well as the final fabricated design. The power output was simulated for each iteration in order to check the deviation from the optimization. Changes from one iteration to the next are marked in bold letters.

Changes in the design include a reduction of the coil width from 10 mm down to 9 mm in order to create 1 mm of space between coils. The circuit board holding the coils can then accommodate the coils in separate pockets. Another change is the reduction of the coil length, again, so that the circuit board can accomodate the coils

K. Ylli and Y. Manoli, *Energy Harvesting for Wearable Sensor Systems*,
Springer Series in Advanced Microelectronics 62,
https://doi.org/10.1007/978-981-33-4448-8_6

Table 6.1 Simulated optimum design of HAC5 in the first column as well as 3 design iterations that arose during the CAD design. The design iteration 6 is the final design, that was fabricated. Parameter changes from one design to the next are shown in bold letters. The simulated power output for each design as well as the power output normalized to 1, i.e. the optimum design, are also given

Parameter	HAC5—second optimization run	HAC5—CAD design iteration 1	HAC5—CAD design iteration 3	HAC5—CAD design iteration 6 (final design)
Magnet height (mm)	2.65	**2.6**	**2.5**	**2.4**
Magnet length (mm)	37.5	37.5	**35**	35
Magnet width (mm)	10.35	**10**	10	10
Coil height (mm)	2.55	**2.5**	2.5	**2.4**
Coil length (mm)	36	36	**32**	32
Coil width (mm)	10	**9**	9	9
Coil inner width (mm)	1	1	1	1
Number of magnets	4	4	4	4
Number of coils	5	5	5	5
Back steel height (mm)	1.8	1.8	1.8	**2**
Magnet stack length (mm)	41.4	**40**	40	40
Simulated P_{RMS} (mW)	22.00	19.22	16.86	17.61
$P_{RMS}/22\,mW$	1	0.87	0.77	0.8

and still fit within the brass brackets that hold the two sides of the harvester together.

In Fig. 6.1 the assembly of the final HAC5 design is shown. One of the components that led to design iterations as explained above is the steel backplate (flux guide).

Although it is very thin, it needs to contain pockets, where the brass brackets are slided in between the steel backplate and the magnets. Without the brackets, the two sides of the harvester would stick together due to the strong attractive forces of the magnetic circuit. The brackets are small and very thin U-shaped structures placed outside the magnetic circuit (the main body of the brass bracket is only 0.5 mm thin), yet need to be strong enough, to withstand the attractive forces. This is one example, where the workshop needed more than one run, to fabricate completely straight shapes (Fig. 6.2).

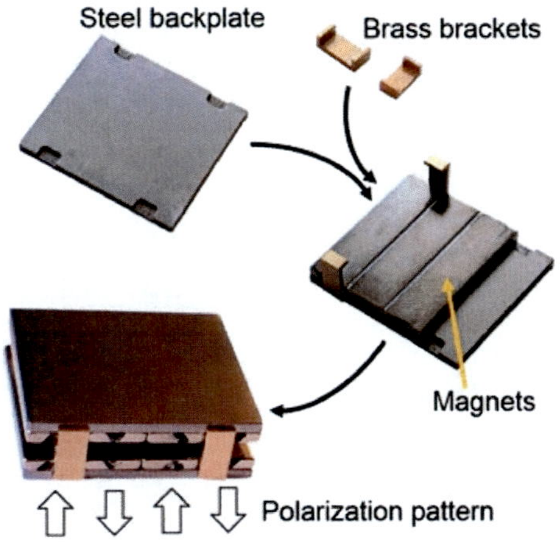

Fig. 6.1 The assembly steps during fabrication of the final HAC5 design are shown

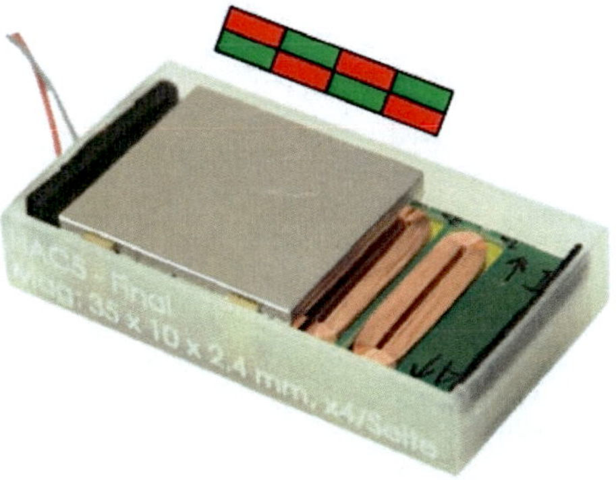

Fig. 6.2 Fabricated second generation of HAC5 based on CAD design iteration 6 in Table 6.1

6.2 Fabrication HAC6

The fabrication of HAC6 follows the same path as described above. From the design generated by the optimization algorithm to the final CAD design that was actually produced, a total of four design iterations were required.

Three of those CAD design iterations are shown in Table 6.2 and again the changes from one iteration to the next are marked in bold letters.

The necessary changes to the optimal design are as follows: The length of the magnets had to be reduced, because a supporting structure is needed to hold the two sides of the Halbach array apart. Other than the length, the other magnet widths had to be modified slightly, because the magnet supplier can only fabricate magnets with a resolution of 0.1 mm. The coil length and width had to be decreased, so that the circuit board can contain single pockets for coil placement, while still fitting within the Halbach array (i.e. within the supporting structure that holds the magnets apart).

The coil placement is done as follows: The PCB is ordered with pockets of the size of a single coil. Thin beams separate the coils. Along the outer margins of the

Table 6.2 Simulated optimum design of HAC6 in the first column as well as 3 design iterations that arose during the CAD design. The design iteration 4 is the final design, that was fabricated. Parameter changes from one design to the next are shown in bold letters. The simulated power output for each design as well as the power output normalized to 1, i.e. the optimum design, are also given

Parameter	HAC6—second optimization run	HAC6—CAD design iteration 1	HAC6—CAD design iteration 3	HAC6—CAD design iteration 4 (final design)
Magnet height (mm)	5	5	**4**	4
Magnet length (mm)	37.5	37.5	**35**	35
Magnet width, 1 (mm)	5.05	**5**	5	5
Magnet width, 2 (mm)	5.45	**5.5**	5.5	5.5
Coil height (mm)	1.25	**1.2**	1.2	1.2
Coil length (mm)	36	36	**33**	**30**
Coil width (mm)	9.25	**9**	9	9
Coil inner width (mm)	1	1	1	1
Number of magnets	7	7	7	7
Number of coils	7	7	7	7
Magnet stack length (mm)	36.55	**36.5**	36.5	36.5
Simulated P_{RMS} (mW)	22.9	21.86	16.3	14.78
$P_{RMS}/22.9\,mW$	1	0.95	0.71	0.65

PCB there are pads to connect the coils and the pads are connected to each other in such a way, that if you connect all coils identically, they are connected in series opposition. Before the coil placement, the whole back side of the PCB is covered carefully with Kapton tape. The coils are then placed within the pockets and soldered onto the PCB with precise and flat soldering points. Afterwards the cold mounting epoxy resin "SpeciFix-20" is used to secure the coils. SpeciFix is a clear epoxy that is easy to handle and can be cured at a comparatively low temperature of 60 °C [1]. After curing, the coils are held in place firmly and the Kapton tape can be removed from the underside of the PCB (Fig. 6.3).

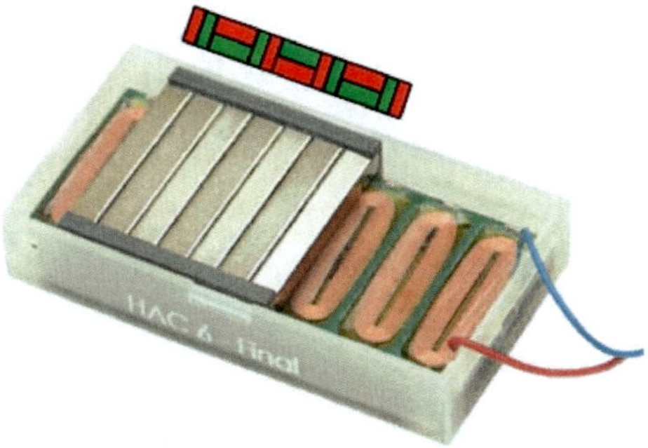

Fig. 6.3 Fabricated second generation of HAC6 based on CAD design iteration 4 in Table 6.2. The top side of the Halbach array is shown, as well as the grey milled POM (Polyoxymethylene) brackets on either side of the magnet stack. The PCB holding the coils fits exactly between the two brackets with a airgap of roughly 0.2 mm on either side

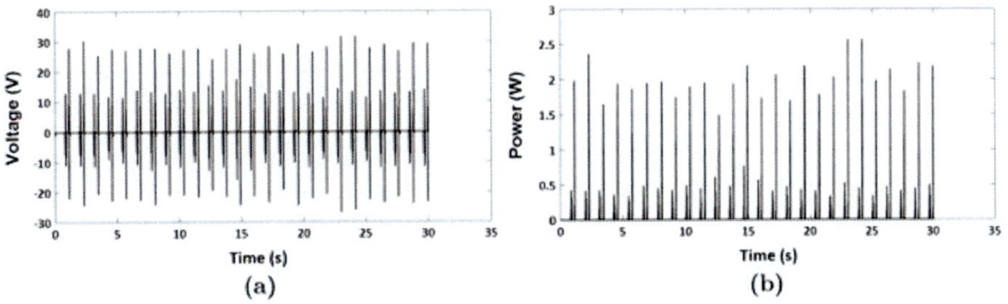

Fig. 6.4 Voltage output across an ohmic load of 390 Ω in (**a**) and the corresponding power output for the final device of HAC5 in (**b**). Measured on a treadmill at a motion speed of 4 km/h

6.3 Experimental Data

6.3.1 HAC5—2. Generation

The experimental setup is identical to the previous experiments described in Sect. 4.4.1. Both the voltage and power output figures greatly surpass the previous generation at voltages up to roughly 30 V and an instantaneous power output of up to 2.5 W at only 4 km/h motion speed (see Fig. 6.4).

Since the used DAQ-cards (data acquisition) can only read voltages up to 10.5 V, a voltage divider with a ratio of 1:2 was used in the measurements. This limits the effective measuring range to approximately 31.5 V. Figure 6.5 shows the voltage and instantaneous power output of the final HAC5 device for a motion speed of 10 km/h.

An average power output of 26.02 mW was measured at 4 km/h and up to 43.16 mW were measured at 10 km/h using an ohmic load of 390 Ω. A detailed overview of the generated power outputs and power densities is given in Sect. 6.4.

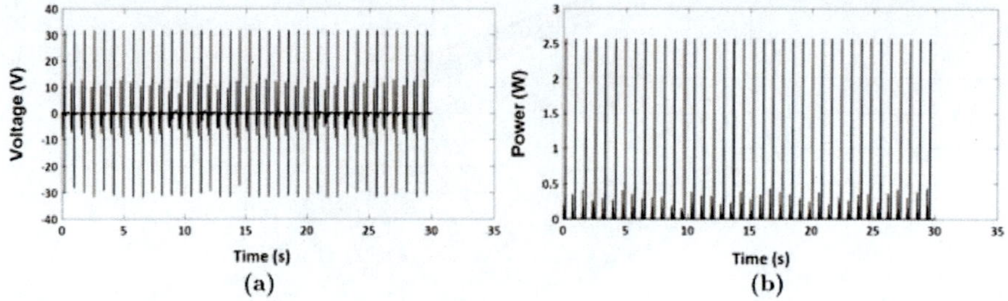

Fig. 6.5 Voltage output across an ohmic load of 390 Ω in (**a**) and the corresponding power output for the final device of HAC5 in (**b**). Measured on a treadmill at a motion speed of 10 km/h

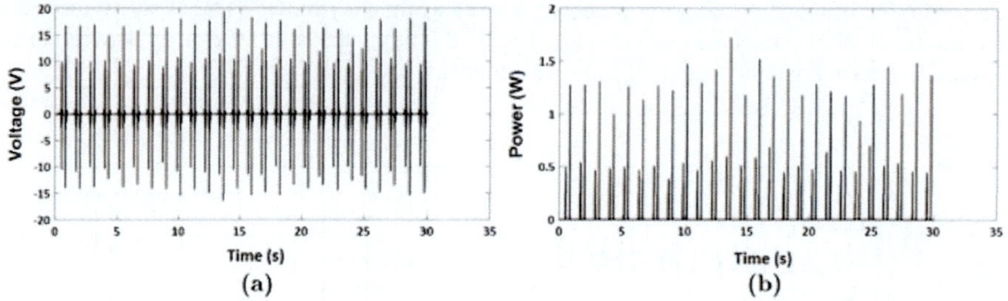

Fig. 6.6 Voltage output across an ohmic load of 220 Ω in (**a**) and the corresponding power output for the final device of HAC6 in (**b**). Measured on a treadmill at a motion speed of 4 km/h

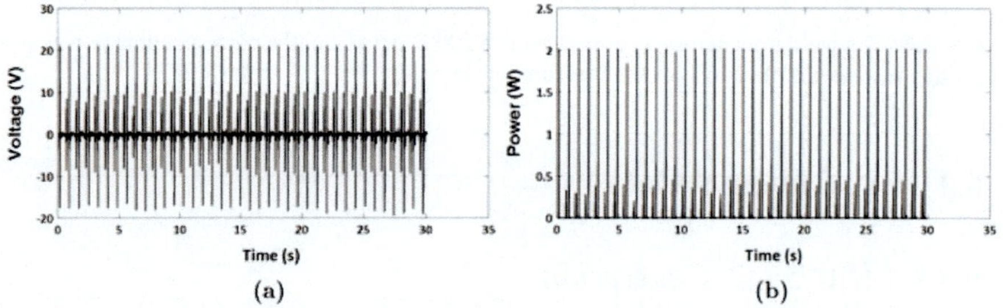

Fig. 6.7 Voltage output across an ohmic load of 220 Ω in (**a**) and the corresponding power output for the final device of HAC6 in (**b**). Measured on a treadmill at a motion speed of 10 km/h

6.3.2 HAC6—2. Generation

In the case of HAC6, the generated voltage and power output is also far larger than for its previous generation device. The generated voltages reach values of up to 18 V while the instantaneous power output peaks at values of up to 1.75 W for a motion speed of 4 km/h (see Fig. 6.6).

For the measurements of HAC6 a voltage divider with a ratio of 1:1 was used. The experimental results at a motion speed of 10 km/h are shown in Fig. 6.7.

For this device an average power output of 19.97 mW was measured at 4 km/h and up to 31.7 mW were measured at 10 km/h using an ohmic load of 220 Ω. These results greatly improve upon the previous generation (see Sect. 6.4).

6.4 Discussion

In Table 6.3 the complete set of experimental results in terms of the average power output of the second generation of harvesters is given. "HAC5-A" and "HAC5-B" denote two fabricated devices of the HAC5 architecture. As was the case in Chap. 4, the experiments were performed with two test persons and for four different motion speeds on a treadmill.

Generally speaking, the highest power output can be generated at the motion speeds of 6 and 10 km/h, analogously to the first generation results. The largest measured power output for HAC5 is 43.16 mW at 6 km/h and for HAC6 it is 31.7 mW at 10 km/h.

When compared directly to the first generation results in Chap. 4, at a motion speed of 4 km/h, "HAC5-B" generates a 363% larger average power output than the best first generation device (5.62 mW for HAC1 XL). Compared to the first generation HAC5 (4.89 mW at 4 km/h), the second generation HAC5 generates a 432% larger average power output. HAC6 shows an even larger improvement of 946% over the first generation device (1.91 mW at 4 km/h).

One particular observation was made concerning the fabrication quality of the devices. HAC5-A and HAC5-B generated similar average power outputs during the runs at 4, 6 and 10 km/h. However, at 8 km/h HAC5-A performs significantly better than HAC5-B. An investigation of the device HAC5-B showed, that the magnetic circuit touches the bottom of the printed circuit board that holds the coils in some positions. This obviously leads to an increased friction. It appears that in the 8 km/h run, the acceleration of the foot did not trigger the magnet motion in several steps, thus leading to a lower power output. This happens, because the foot motion at a speed of 8 km/h (slow jogging) is very smooth and does not contain acceleration peaks as sharp as is the case for the other motion speeds considered in this work.

Another problem occured in the second set of experiments with HAC5-A as shown in the bottom part of Table 6.3. In this case, due to a slight misalignment of the circuit board after a number of experiments, the Nickel-coated magnets could touch the points, where the coils are soldered onto the circuit board and thus short-circuit the coils. The consequence is a drastic drop in power output, for example at 10 km/h. In future work, the air gaps between the components should be kept slightly larger, to avoid complications due to fabrication tolerances or slight misalignments.

The reader may have noticed the difference between the value the simulated power output given in Table 6.3 and the values given for the final design iterations in Tables 6.1 and 6.2. In Table 6.3 the simulated power output is larger than in the previous tables. That difference is caused by two reasons. The first reason is, that in the optimization runs a magnetization of N35 is assumed in the simulations, as

Table 6.3 Overview of all performed measurements and the average power output that was generated in each case. "1" and "2" denote the test persons. Measurements were performed at four motion speeds. For the motion speed of four 4 km/h and test person "1", the simulated average power output is also listed. The bottom part of the table shows data from a second measurement set with the HAC5 devices

	4 km/h			6 km/h		8 km/h		10 km/h	
	"1"	Sim "1"	"2"	"1"	"2"	"1"	"2"	"1"	"2"
HAC5-A (mW)	25.33	21.42	19.54	43.16	26.43	33.20	18.82	40.80	24.72
HAC5-B (mW)	26.02	21.42	20.84	40.62	29.62	18.72	11.87	42.10	29.59
HAC6 (mW)	19.97	17.10	17.86	30.58	24.30	24.24	25.38	31.70	29.99
HAC5-A (mW)	22.87	21.42	22.19	38.39	32.77	0.28	14.91	6.16	5.09
HAC5-B (mW)	22.98	21.42	22.87	42.25	30.24	30.34	30.20	41.81	34.37

that is the most common magnetization and widely available. Once the design of the magnetic circuit was clear, quotes could be obtained for stronger magnetizations. In the case of the fabricated second generation devices, magnets with a magnetization of N52 were used for HAC5 and magnets with a magnetization of N48 for HAC6. The second reason is, that the custom-wound air coils were better than expected and contained an increased number of windings. Once these two values were adapted in the simulation, the simulated power output increased to the values given in Table 6.3.

The deviation between simulation and experiment reaches its maximum of 21% for HAC5-B, while it is 17% for HAC6. However, as was mentioned in previous chapters, there are quite large irregularities in the human gait, although the person might not notice. Therefore a second set of experiments was performed for the HAC5 devices as shown in the bottom part of Table 6.3. Here, in the case of HAC5-A, the deviation between simulation and experiment is only 7%. It can be summarized, that the simulation is a good indicator of the kind of power output that can be expected from such a device, but it is far from a precise prediction due to the irregularities in the gait. Based on the experience gained during this work, deviations of 10% should always be expected.

The power densities that were calculated from the experimental results in Table 6.3 are given in Table 6.4 with the respect to the active (transducer) volume of 34.22 cm^3 and in Table 6.5 with respect to the total volume including the housing of 50.33 cm^3.

As was the case for the power output, the largest power densities can be found for fast walking at 6 km/h and fast jogging at 10 km/h. Looking at the motion speed

Table 6.4 Overview of all the power densities corresponding to the top part of Table 6.3 for an "active" volume of 34.22 cm^3. "1" and "2" denote the test persons

Power densities for an active volume (transducer volume) of 34.22 cm^3 in mW/cm^3								
	4 km/h		6 km/h		8 km/h		10 km/h	
	"1"	"2"	"1"	"2"	"1"	"2"	"1"	"2"
HAC5-A	0.74	0.57	1.26	0.77	0.97	0.55	1.19	0.72
HAC5-B	0.76	0.61	1.19	0.87	0.55	0.35	1.23	0.86
HAC6	0.58	0.52	0.89	0.71	0.71	0.74	0.93	0.88

Table 6.5 Overview of all the power densities corresponding to the top part of Table 6.3 for a total volume including housing of 50.33 cm^3. "1" and "2" denote the test persons

Power densities for a total volume (incl. housing) of 50.33 cm^3 in mW/cm^3								
	4 km/h		6 km/h		8 km/h		10 km/h	
	"1"	"2"	"1"	"2"	"1"	"2"	"1"	"2"
HAC5-A	0.50	0.39	0.86	0.53	0.66	0.37	0.81	0.49
HAC5-B	0.52	0.41	0.81	0.59	0.37	0.24	0.84	0.59
HAC6	0.40	0.35	0.61	0.48	0.48	0.50	0.63	0.60

of 6 km/h, HAC5-A generates its largest average power density of 1.26 mW/cm^3. That represents a 23% increase over the best first generation result (1.03 mW/cm^3 for HAC1 XL at 6 km/h). This is a relatively moderate improvement and can be explained by the fact, that the cylindrical setup of HAC1 makes a good use of the transducer volume. However, the width of the housing cannot easily be used for parallel transducers of HAC1 due to magnetic forces as was explained in Chap. 4. Therefore the overall power output of a flat device based on HAC1 is quite limited.

When the above power density of 1.26 mW/cm^3 is compared to the first generation device of HAC5, the improvement is 434% and identical to the improvement in power output, apart from small rounding errors. HAC6 reaches an average power density of up to 0.89 mW/cm^3 at 6 km/h which is a 522% increase over the first generation device. This is a smaller improvement than for the power output, which can be explained by the increased transducer volume of the second generation HAC6.

Looking at the power density for the total volume including the housing, HAC5-A reaches it largest average power density of 0.86 mW/cm^3 at 6 km/h. This value represents an 87% increase over the best first generation result (0.46 mW/cm^3 for HAC1 XL at 6 km/h, assuming a narrow housing that can house only one cylindrical transducer). Compared to the first generation HAC5 device, the improvement amounts to 438% (0.16 mW/cm^3 for the first generation HAC5 at 6 km/h).

HAC6 also shows a great improvement of 648% over the first generation device (0.08 mW/cm^3 for the first generation HAC6 at 6 km/h). Although the transducer volume and total volume have grown compared to the previous generation of HAC6, the power density has increased dramatically, confirming the improvement of the magnetic circuit.

Altogether these new results show, that a full system optimization can be used to find the best possible harvester geometry for the application in the shoe to harvest energy during walking. In this chapter, very good experimental results with average power outputs of upto 43 mW were achieved for flat, non-resonant inductive harvesters, simply by improving the system model and confirming the implementations through experiments.

Reference

1. Struers GmbH, Anwendungshilfe kalteinbetten: https://cdnstruersproduction.azureedge.net/-/media/struers-media-library/products/mounting/mounting-consumables/selection-guide-for-cold-mounting-de.pdf?la=de-de, 22 Dec 2018

Chapter 7
Industrial Applications

Linear accelerations that can be used as an excitation source for the kind of harvester described in this book can not only be found in human motion, but obviously also in many industrial applications. This chapter presents two suitable applications. The first application is a pneumatic piston, which is not only used to prove that the HAC devices work in non-human applications, but also to compare the first generation HAC devices from Sect. 4.5 with each other. The human gait is a motion type that contains many irregularities. As was shown in previous chapters, when performing measurements with test persons, even on a treadmill that guarantees a controlled pace, the accelerations vary and so do the power outputs (see Sect. 4.5 and Chap. 6). In order to be able to correctly compare the harvesting devices, it is possible to use a large number of test persons and determine statistically relevant averages in power output. The second approach for comparing the devices is to use the aforementioned pneumatic piston as a regulated, non-human excitation source. The setup of this approach is presented in Sect. 7.1 and the experimental results are presented in Sect. 7.2.

An outlook on a second industrial application, clutch-brakes, is given in Sects. 7.3 and 7.4. These mechanical machines are used to transmit large amounts of energy from a rotating part onto another mechanical device, e.g. a press, in a very short time. An energy harvesting device based on the HAC5 architecture was developed for this application. The operating principle of a clutch-brake and the experimental setup are described in Sect. 7.3. The first set of experimental results is given in Sect. 7.4.

7.1 Pneumatic Piston—Setup

The following sections describe the use of a piston as an energy source in a laboratory environment. The piston is a pneumatic piston of type "DSBC-32-150-PPVA-N3" from "Festo". The air pressure to the device can be restricted to change the acceleration and speed of the piston. It has a total motion range of 150 mm.

K. Ylli and Y. Manoli, *Energy Harvesting for Wearable Sensor Systems*,
Springer Series in Advanced Microelectronics 62,
https://doi.org/10.1007/978-981-33-4448-8_7

7.1.1 Experimental Setup

The pneumatic piston used in this measurement setup can be operated at up to 12 bar of pressure, however, here it is operated at the standard pressure of 6 bar. Depending on the load (weight), the maximum speed that is achieved by the piston within its 150 mm stroke is approximately 2 m/s. Chokes are used in the measurement setup, which halve the pressure to the piston and reduce its maximum speed to approximately 1 m/s.

All relevant components of the measurement setup are depicted in Fig. 7.1. The piston controller receives a voltage signal from a laboratory power supply (24V, 1A) and switches a valve that directs the air pressure towards either end of the piston. The power supply as well as the data acquisition card (National Instruments N9205 placed in an NI cDAQ-9172 rack) are controlled through a customized LabView program. The harvesting devices can be placed in a sledge at the tip of the piston rod and perform the full motion along the stroke of the piston. A three axis acceleration sensor type "52413" from "Bruel & Kjaer" with a 50 g measurement range is used to set up the piston before the harvesters are characterized (see Sect. 7.1.2). The accelerations are not measured when a harvester is attached as the sensor also records the impacts of the magnet stack with the end stops which show as large spikes (beyond 50 g) in the acceleration data and falsify the remaining signal. The sampling rate of the acceleration measurements is 25.6 kilosamples per second, giving a good resolution of the narrow acceleration peaks.

One full motion cycle consists of the forward and backward motion of the piston. The two motion phases are inherently asymmetric in terms of acceleration and speed because in one case the pressure can act on the full cross-section of the piston, while in the other case, the "active area" of the cross-section is reduced by the piston rod. Two chokes are implemented in the setup, which can be used to separately control the pressure available for each of the motion phases. The piston can thus be regulated to show the same accelerations in both forward and backward motion.

Every measurement is performed along a period of 40 s. During this time period a total of 30 motion cycles of the piston is carried out. The voltages generated by the harvesters are measured across ohmic loads as used in Table 4.5. The data is sampled at 5000 samples per second leading to a total of 200000 data points per measurement.

7.1.2 Comparison to Human Gait

This section is dedicated to a brief comparison of the accelerations that occur at the shoe during walking and at the piston. Obviously, the stroke of the piston (150 mm) is far shorter than the normal human step size. During this phase, the piston accelerates to its maximum speed and decelerates down to a standstill. In Fig. 7.2 acceleration graphs are given for the walking motion recorded at 4 km/h on a treadmill as well as for a number of different conditions the piston can be operated in.

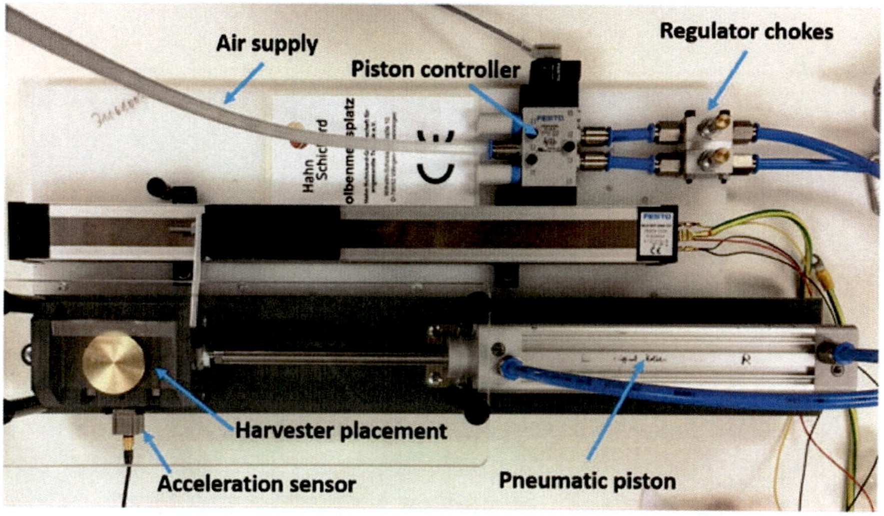

(a)

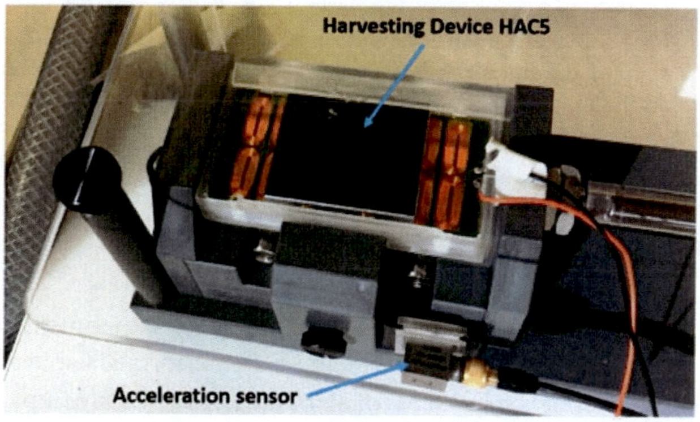

(b)

Fig. 7.1 The measurement setup with a pneumatic piston and the required components is shown in (**a**). The harvester placement, in this case device HAC5 (1st generation), is shown in more detail in (**b**). Figures from [3]

Immediately the difference in the duration of the acceleration impulses can be noticed. The walking motion generates a comparatively wide acceleration peak as shown in Fig. 7.2a. When the piston is used, the peaks are far narrower, particularly in the unregulated case in Fig. 7.2b. In this case the chokes are set to their maximum setting and the piston reaches accelerations up to 17g. The chokes settings are then changed to reduce the magnitude of the accelerations down to values that approximately match the acceleration that occur during the leg motion. Figure 7.2c shows the case where the accelerations are reduced to 4–5 g while in Fig. 7.2d the acceleration peaks out at about 7g. This is not a highly exact setting and the peaks within the signal show some variation.

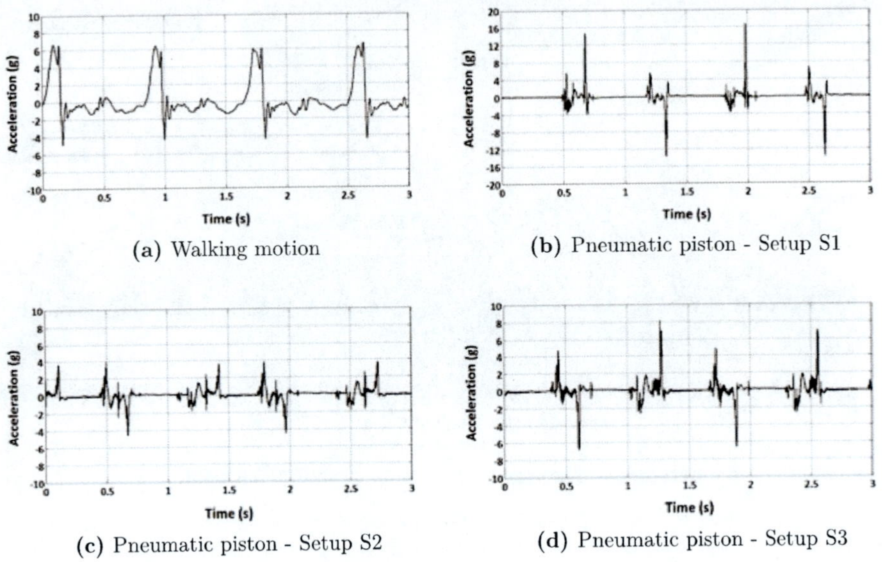

(a) Walking motion

(b) Pneumatic piston - Setup S1

(c) Pneumatic piston - Setup S2

(d) Pneumatic piston - Setup S3

Fig. 7.2 Comparison of Accelerations: Walking on a treadmill at 4 km/h in (**a**), the motion of the unregulated piston in (**b**), the motion of the piston with pressure regulated to produce accelerations of 4–5g in (**c**) and the motion of the piston regulated to produce accelerations of 6–7g in (**d**). Figures from [3]

7.2 Pneumatic Piston—Experimental Results

In the following, the measurement results for the piston experiments are presented. The load sweeps for each harvester architecture are performed using the unregulated piston. The power output is measured for all of the above piston setups and each measurement is repeated five times. "S17" indicates measurements with the unregulated piston where the accelerations reach up to 17g and this shows in the power output figures, where "S5" indicates measurements where the piston is regulated to peak out at 5g and "S7" indicates measurements where the piston peaks out at approximately 7g. All the following experiments were conducted with the first generation of "HAC" harvesters described in Chap. 4.

7.2.1 HAC1—1. Generation

The harvesting devices are mounted to the sledge at the tip of the piston rod one by one and the measurements are performed, leading to the power outputs given in Table 7.1. The setup S17 leads to the largest power output. This is the expected result, as the accelerations in S17 are far larger than in S5 and S7. The power outputs follow

Table 7.1 The average power output of HAC1 is given for all measurements and each of the three piston settings. Table from [3]

	Measurement, power output (mW)					
	1	2	3	4	5	Average
S17	3.30	3.43	3.35	3.50	3.57	3.43
S5	1.33	1.33	1.33	1.34	1.35	1.34
S7	1.83	1.82	1.80	1.80	1.78	1.81

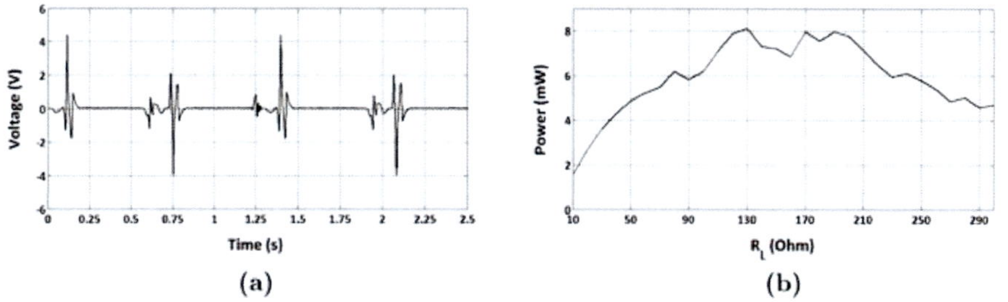

Fig. 7.3 The voltage signal of HAC1 measured across the ohmic load of 70 Ω and piston setup S17 is given in (**a**). The load sweep for HAC1 is shown in (**b**) (the total coil resistance of HAC1 is 70 Ω). Figures from [3]

the magnitude of the accelerations of the piston and rank accordingly with HAC1 generating more power under condition S7 than under condition S5.

Figure 7.3 shows the voltage signal across a load equal to the coil resistance (70 Ω) as well as the load sweep. The data for both graphs was measured under piston setup S17. The voltage signal appears stable and the signal pattern is repeated regularly. The following sections of this chapter will show, that this observation does not apply to all harvester architectures investigated herein. A repeated voltage pattern with voltage peaks of the same magnitude implies, that the motion of the magnet stack is regular, i.e. that the stack comes to rest in nearly the same position in each rest phase of the piston. As a new motion phase begins, the starting conditions are comparable. This property of the magnet stack motion was also observed during treadmill runs, where the magnet appeared to come to rest in the same position during each stance phase.

The generated voltage using the piston reaches values of roughly 4 V, which compares well to the voltages measured on a treadmill run at 4 km/h. However, the voltage pulse is noticeably narrower, thus explaining the lower power outputs compared to the treadmill measurement (see Sect. 4.4.2).

The load sweep shows some fluctuation, but overall it can be noticed that the optimum load can likely be found on a plateau between 130 Ω and 200 Ω. This is a larger value than the one found during the load sweeps performed on the treadmill, where the optimum load was determined to be roughly 70 Ω. This effect can likely

Table 7.2 The average power output of HAC1 XL is given for all measurements and each of the three piston settings. Table from [3]

	Measurement, power output (mW)					
	1	2	3	4	5	Average
S17	6.56	6.43	6.72	6.68	6.71	6.62
S5	2.48	2.53	2.53	2.55	2.54	2.53
S7	3.42	3.30	3.42	3.41	3.45	3.40

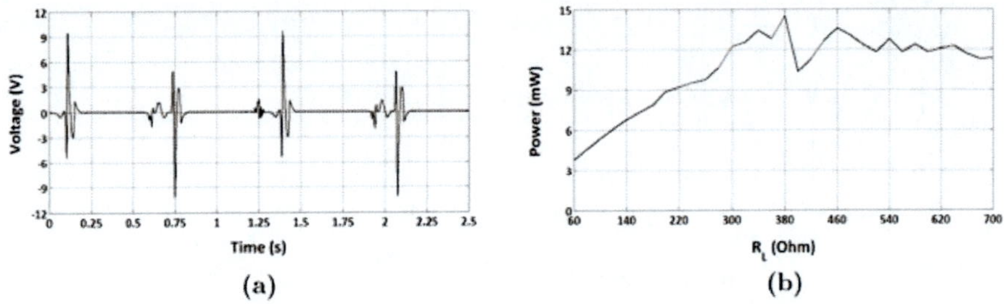

Fig. 7.4 The voltage signal of HAC1 XL measured across the ohmic load of 220 Ω and piston setup S1 is given in (**a**). The load sweep for HAC1 XL is shown in (**b**) (the total coil resistance of HAC1 XL is 220 Ω). Figures from [3]

be attributed to the irregularities in the excitation pattern, as it cannot be observed in the case of the remaining architectures.

Considering HAC1 XL, the piston setup S17 generates the largest power output (Table 7.2) which can be attributed to the large accelerations of the piston. The voltage signal measured across an ohmic load of 220 Ω in Fig. 7.4 also appears to show a regular pattern and the voltages reach higher values than was the case in the treadmill runs (see Sect. 4.4.2).

Due to the high voltages, the power output is correspondingly high and HAC1 XL manages to generate about twice the power of the standard HAC1 in setup S17.

The load sweep in Fig. 7.4 shows a jump in the power output between 380 Ω and 400 Ω. Other than that, the optimum load is found at roughly 380 Ω. This corresponds well to optimal load found in Sect. 4.4.2.

In the piston measurements the shape of the graph is more pronounced, as opposed to the rather flat shape generated during treadmill measurements. This gives a clearer indication of where the optimal load is located.

7.2.2 *HAC3—1. Generation*

The third device under investigation behaves as before, with the most power being generated using piston setup S17 (see Table 7.3). The power output is larger than

Table 7.3 The average power output of HAC3 is given for all measurements and each of the three piston settings. Table from [3]

	Measurement, power output (mW)					
	1	2	3	4	5	Average
S17	0.90	0.93	0.92	0.93	0.94	0.92
S5	0.57	0.57	0.55	0.55	0.56	0.56
S7	0.60	0.57	0.58	0.57	0.64	0.59

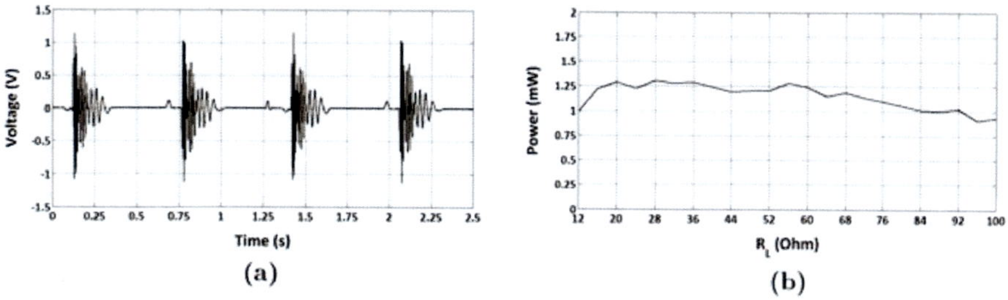

Fig. 7.5 The voltage signal of HAC3 measured across the ohmic load of 33 Ω and piston setup S17 is given in (**a**). The load sweep for HAC3 is shown in (**b**) (the total coil resistance of HAC3 is 33 Ω). Figures from [3]

during treadmill runs (see Sect. 4.4.3). The generated voltage in Fig. 7.5 is measured using an ohmic load of 33 Ω and reaches values up to about 1 V.

Since the number of magnets used in HAC3 is larger than in HAC1, the voltage signal shows more consecutive peaks for each motion phase. The signal in Fig. 7.5 is relatively noisy at the beginning of each motion phase (i.e. at the time of the first large voltage peaks), compared to Figs. 7.3 and 7.4. This may indicate issues with static friction, e.g. through a small misalignment between magnets or steel spacers, that affect the motion.

The load sweep resulted in a rather flat curve as shown in Fig. 7.5. This curve is very similar to the one obtained on a treadmill and the optimal load is located on a plateau between 20 Ω and 40 Ω and cannot be distinguished more clearly using the available data.

7.2.3 HAC5—1. Generation

The generated power of the HAC5 device with piston setup S17 tops all the other architectures at an average of 13.35 mW (see Table 7.4). This is almost three times the power output obtained on a treadmill at 4 km/h (see Sect. 4.4.4). There is a noticeable fluctuation in the power output of the five performed measurements. For the previous architectures, this effect has been small and negligible.

Table 7.4 The average power output of HAC5 is given for all measurements and each of the three piston settings. Table from [3]

	Measurement, power output (mW)					
	1	2	3	4	5	Average
S17	12.14	12.56	13.89	14.05	14.09	13.35
S5	2.98	2.97	2.95	2.86	2.87	2.93
S7	3.68	3.73	3.96	4.16	4.29	3.96

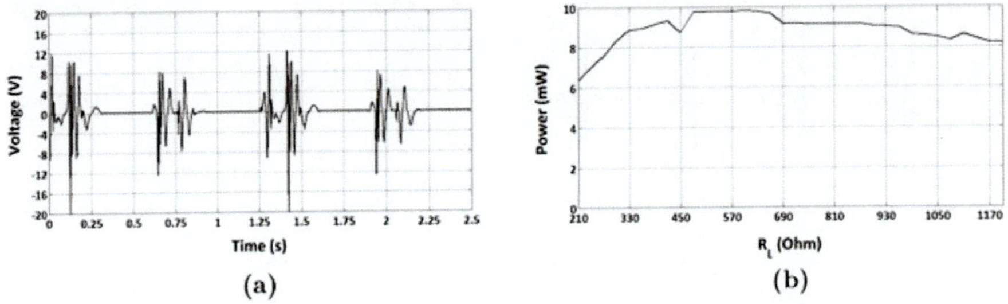

Fig. 7.6 The voltage signal of HAC5 measured across the ohmic load of 480 Ω and piston setup S17 is given in (**a**). The load sweep for HAC5 is shown in (**b**) (the total coil resistance of HAC5 is 480 Ω). Figures from [3]

A slow-motion video (recorded at 240 frames per second and replayed at a tenth of that frame rate) was used to check if there are any recognizable irregularities in the motion of the magnet stack. Indeed, the magnet stack often appears to come to rest in a different spot along the length of the housing after the motion phase of the piston. This changes the starting condition for the next motion phase and thus influences the generated voltage and power. The acceleration impulses are strong enough for the heavy magnet setup to cross the housing several times, bouncing off the rubber hard stops in a rather strong way. It is unlikely that a 3D-printed housing could withstand this kind of impact, without the supporting structures that surround it in the piston setup. The additional version HAC5F which contained plastic springs was not characterized on the piston setup, because the springs were not able to withstand the mechanical load. Several springs broke during testing.

The voltage signal in Fig. 7.6a shows peaks that reach approximately 20 V. When there are two groups of voltage peaks in close proximity, it indicates that the magnet stack has reached the end point and still has enough kinetic energy to perform a motion in the opposing direction. The load sweep for HAC5 generated a comparatively flat curve as shown in Fig. 7.6b. This is a similar result to the treadmill measurements in Sect. 4.4.4. The optimal load cannot be distinguished clearly and is located roughly between 480 Ω and 660 Ω.

Table 7.5 The average power output of HAC6 is given for all measurements and each of the three piston settings. Table from [3]

	Measurement, power output (mW)					
	1	2	3	4	5	Average
S17	8.68	7.36	7.12	7.45	7.06	7.53
S5	3.01	3.21	2.84	2.75	2.91	2.94
S7	4.60	4.49	4.36	4.46	4.44	4.47

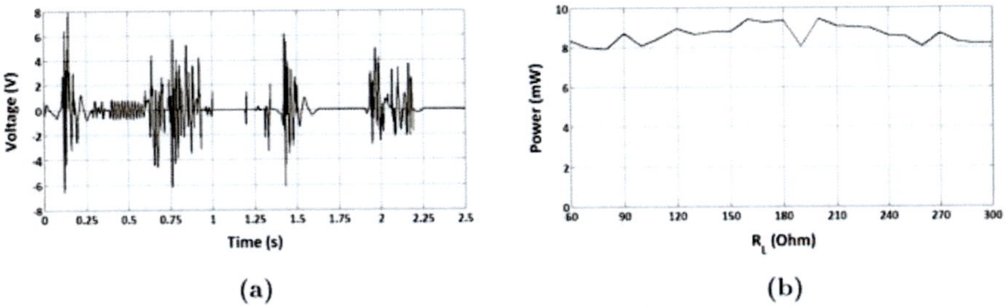

Fig. 7.7 The voltage signal of HAC6 measured across the ohmic load of 118 Ω and piston setup S17 is given in (**a**). The load sweep for HAC6 is shown in (**b**) (the total coil resistance of HAC6 is 118 Ω). Figures from [3]

7.2.4 *HAC6—1. Generation*

The power output generated by HAC6 in piston setup S17 is 7.53 mW on average and almost four times the power generated during treadmill runs at 4 km/h. Table 7.5 shows large fluctuations in the power output. Similarly to HAC5, the magnet stack changes its motion pattern frequently.

This can be observed quite clearly in the very noisy and irregular voltage signal in Fig. 7.7. There is no recognizable and repeated pattern in the signal, unlike e.g. in the case of HAC1. Using a slow-motion video, it is possible to see that there are several phases within one motion of the piston, where the magnet stack appears to be quickly transitioning between static and kinetic friction. This is the case when the piston has reached its maximum speed and does not accelerate any further before decelerating.

The kinetic energy and inertia of the magnet stack cause the small shifts in the magnet position which in turn cause the jitter in the voltage signal. As soon as the piston begins to decelerate, the magnets move across the housing fully and are reflected by the rubber end stops several times, causing the cluster of narrow peaks in the voltage signal.

One possible explanation for the noisy section of the signal between 0.25 s and 0.6 s is the end-stop behavior of the piston itself. Within its housing a plastic cone slows down the piston very quickly, so it does not hit the mechanical end stop at speed.

However, once the cone is entered by the piston rod, the air pressure keeps it moving up to the mechanical hardstop. This is a very brief motion with a duration of less than a second and barely noticeable without technical aid, where vibrations may be induced into the system (the acceleration measurements of S5 in Fig. 7.2c and S7 in Fig. 7.2d also show noisy sections). A parallel measurement of both acceleration and generated voltage however is not possible, as the highly sensitive accelerometers would be influenced by the impact of the magnet stack at the end stops.

7.2.5 Discussion

The power outputs and the corresponding power density figures for all devices are summarized in Table 7.6 for the piston setup S17. Interestingly, the order from the treadmill measurements has changed (compare Table 4.8). In the laboratory experiments, HAC5 generated the most power by a fair margin and HAC6 was able to generate the second-most power.

A factor to be considered in this discussion is the weight of the moving parts, i.e. of the magnet stacks. The rectangular devices, HAC5 in particular, feature far heavier magnet stacks than the cylindrical devices. This translates into a larger normal force and thus into a larger friction force.

The excitation of the harvesters during normal walking generates comparatively small and wide peaks, as is shown in Fig. 7.2, whereas the piston generates a stronger excitation. It can safely be assumed, that the quick motion of the piston suffices to quickly overcome the static friction of the devices and the large accelerations give the magnet stacks enough kinetic energy to reflect from the end stops and cross the housing several times.

This effect was not observed experimentally during walking, where the magnet stacks quickly came to rest after a single impact with the end stops. Due to the larger accelerations at hand, the heavy architectures, e.g. HAC5 at almost 100 g, can be excited strongly by the piston and are able to generate far more electrical power than was the case during walking, while the comparatively light HAC1 (at approx. 20 g) does not show an increase in performance.

The power density for both rectangular devices is not as large as can be achieved with HAC1 XL. However, at this point the results from prior treadmill experiments in Chap. 4 and the second optimization run in Chap. 5 have to be considered. It was shown in Table 4.5, that revisions to the system model were required in order to achieve a good match between the experiments and the simulations. This in turn means, that all fabricated devices are based on geometrical parameters that are not optimal for the given volume.

The new optimization run proved this point. When analyzing the second optimization run and the resulting devices, it was found that HAC1 is already close to the new optimum in terms of power output. The remaining architectures on the other hand, are far off from the optimal power output. HAC5 and HAC6 for instance, have significant potential to surpass the power densities achieved with HAC1.

Table 7.6 Power output and power densities of all architectures for laboratory measurements using a pneumatic piston and the above described measurement setup S17. In the case of cylindrical architectures HAC1 and HAC3, two complete magnet stacks fit within the housing. In an actual device, the power output may be lower depending on the attractive force between magnet stacks. Table from [3]

	HAC1	HAC1 XL	HAC3	HAC5 w/o springs	HAC6
Power output (mW)	3.43	6.62	0.92	13.35	7.53
Power output (mW) (two stacks per housing)	6.86	13.24	1.84	–	–
Total volume (cm^3) (one stack/two stacks)	20.01 / 50.33	20.01 / 50.33	20.01 / 50.33	50.33 /-	40.02 /-
Power density ($\mu W/cm^3$) (one stack/two stacks)	171.4 / 136.3	330.8 / 263.1	46.0 / 36.6	265.2 /-	188.2 /-
Active volume (cm^3) (one stack)	8.96	8.96	8.96	34.22	22.81
Power density ($\mu W/cm^3$)	382.8	738.8	102.7	390.1	330.1

Table 7.7 Comparison of experimental and simulated power outputs of the HACs using piston setups S5 and S7 which show accelerations in the same ranges as they appear in human leg motion. Table from [3]

	Setup S5 (mW)		Setup S7 (mW)	
Device	Experiment	Simulation	Experiment	Simulation
HAC1	1.34	2.48	1.81	3.73
HAC1 XL	2.53	2.57	3.40	4.93
HAC3	0.56	0.29	0.59	0.58
HAC5	2.93	4.17	3.96	9.19
HAC6	2.94	1.10	4.47	2.55

The above discussion only considered experimental results. However, a key component of this work is the system model. Table 7.7 lists the experimental power outputs against the simulated power outputs for the two piston setups S5 and S7 which show accelerations with magnitudes that appear in human leg motion. A general tendency cannot be observed in the deviations.

The simulated power output is far larger than the experimental value for HAC1, HAC1 XL and HAC5. For the remaining architectures the simulated values are smaller than the experimental values. This means, that there is no correlation between the weight of the magnet stack and deviation between simulation and experiment.

It was mentioned previously, that the magnet stacks of the rectangular architectures HAC5 and HAC6 are heavier than is the case for the cylindrical architectures HAC1 and HAC3.

The simulated power output however, is larger than measured for HAC5 and smaller than measured for HAC6. The difference between simulation and experiment is large in all cases except HAC1 XL for S5 and HAC3 for S7. The following paragraphs explain, why the system model is not suitable for these very large and narrow acceleration peaks. It is assumed that these two close matches (out of 10 measurement sets) are a coincidence, as the deviation between simulation and experiment is very large for example for the second measurement set of HAC1 XL with setup S7.

In the simulations, the acceleration signal from the piston motion is used as the excitation source. The signal was filtered before being used in Simulink. The noise in the signal, particularly around zero, causes a large number of zero-crossings in the simulation whenever the signal changes its sign. Zero-crossings are detected as special or "important" events by the simulation environment and this leads to an increased simulation time and in the worst case to the abortion of the simulation as the simulation environment cannot cope with a large number of zero-crossings. The signal was filtered in Matlab using a second-order Butterworth filter with a cutoff frequency 50 Hz. Figure 7.8a shows an enlarged section of both the original (black line) and the filtered acceleration signal (red line). As was mentioned before, the acceleration impulses on the piston are extremely narrow with durations down to approximately 10 ms. The duration of acceleration impulses is far shorter than is the case with human motion, where acceleration peaks can have a duration of more than 100 ms (see Fig. 7.2).

In Chap. 2 the system model was explained and it was shown that it contained feedback loops which return the forces acting on the magnet stack for the next simulation step. This means, that there is no instantaneous reaction to a change in one of the relevant forces. One simulation step always passes due to the nature of the discrete numerical simulation, before the integrator blocks can react to a change in their input and thus recalculate the magnet position and speed, consequently leading to new values for the relevant forces.

In addition to this, a direct feedback of the forces back to the integrator block is not possible as it causes an algebraic loop which Simulink's internal algorithms cannot resolve. This is best explained at the start of a simulation where a force value is required at the integrator input, but no value exists at this moment. Thus an initial condition is used (magnet position is set to the middle of the harvester and the magnet speed is set to zero) with a so-called "memory" block (small white block in the bottom right of Fig. 2.3), to break the algebraic loop by providing the data from the integrator output to the onward data stream with an additional delay

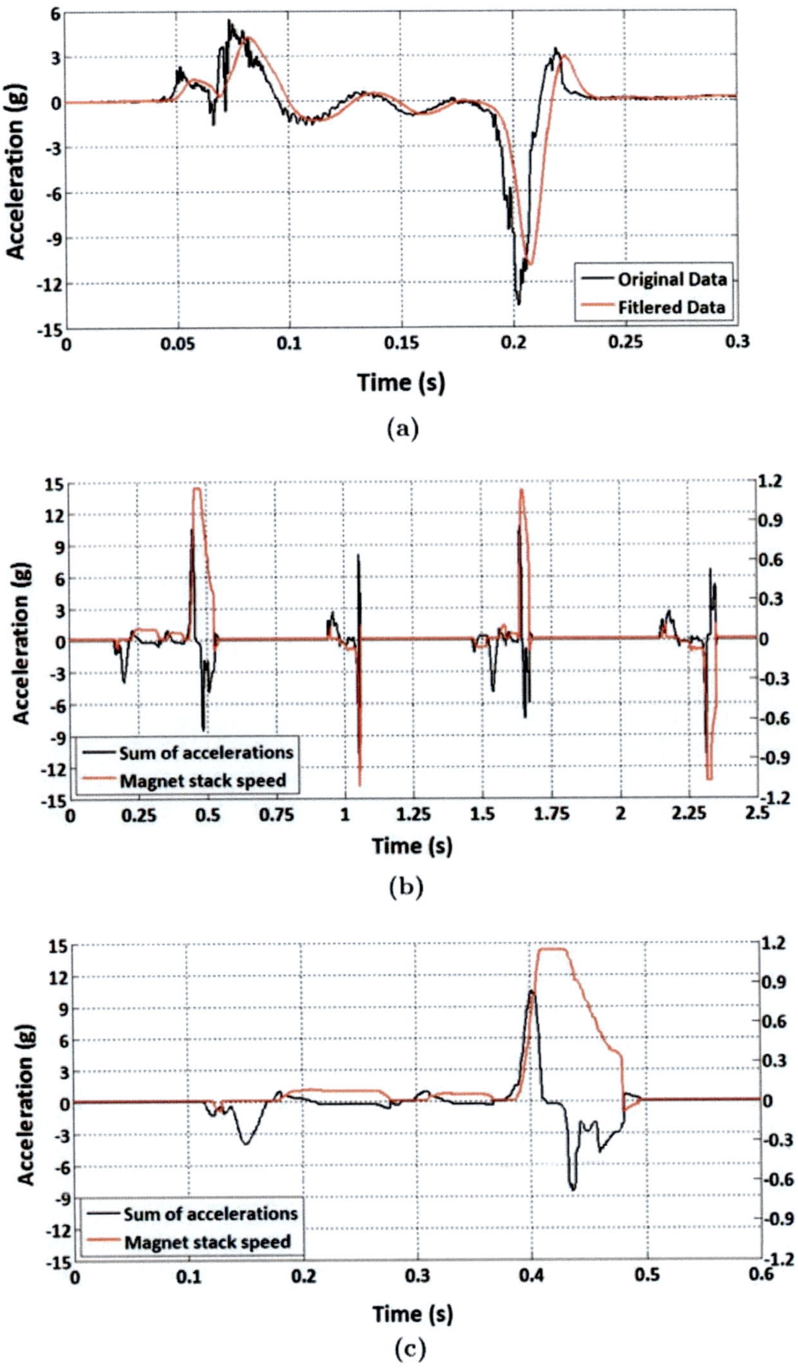

Fig. 7.8 The measured as well as the filtered accelerations for piston setup S17 are given in (**a**). In (**b**) the acceleration input into the integrator block in Simulink is shown in black and the resulting simulated magnet stack speed in red. (**c**) Shows an enlarged section of the signals in (**b**). Figures from [3]

of a simulation step. Basically, a total of two simulation steps of delay pass between action and reaction in the system model.

This effect is indicated in Fig. 7.8b, which shows the acceleration input to the integrator block in the Simulink model (black line) and the resulting magnet stack speed (red line). The acceleration signal is generated from all the forces acting on the magnet stack, i.e. external acceleration, friction and electrical damping because of the voltage generated in the coils. An enlarged section of this signal can be seen in Fig. 7.8c. When analyzing the peak in the speed graph starting at 0.4 s, it can be observed, that the rate at which the speed of the magnet stack increases, does not decrease although the accelerations show a large decrease down to zero. Here the speed signal lags behind in its reaction to the acceleration signal. When the acceleration signal contains peaks of longer duration, as is the case with human motion, the small lags (on the order of 1 μs to 1000 μs because of the use of a variable-step solver) within the system model can hardly be noticed in the power output.

With faster acceleration peaks, this lag can have a significant effect. It is possible to reduce the simulation step size, or set it to a fixed value, but this comes at a significantly increased computation time (using a maximum time step of 10 μs instead of 1000 μs had only a small effect on the power output, but increased the simulation time from approximately 40 s up to 26 min). Additionally, a finer resolution of the simulation and thus the magnet motion does not relieve the model of uncertainties concerning the real-world friction behavior and hard-stop behavior, especially at large speeds of the magnet stack.

From all simulations performed thus far, it must be concluded, that the simulation model presented in this work is not able to calculate the correct power output of a harvesting device, when the acceleration data contains large and narrow peaks. The simulation cannot properly follow those large and abrupt changes in its input. The experiments however proved, that this harvester type can generate large power outputs in industrial applications. Therefore, a detailed investigation of suitable solvers, simulation configurations and the behavior of stiff mechanical systems should be performed in future work, in order to apply the system model to the shown excitation types.

7.3 Clutch-Brake—Setup

In the following, the industrial application of clutch-brakes is described. A device based on the HAC5 architecture was developed for this application [1]. However, at the time of the project the system model was not finished yet, so the device dimensions are based on experience and not fully optimized as was the case for the second generation of harvesters shown in Chap. 6. Nevertheless this is a very promising application for that type of inductive harvester and the first set of experimental results is given in Sect. 7.4.

Clutch-brakes are devices that are used to transmit large amounts of energy. A heavy metallic fly-wheel is set into motion using a powerful electric motor. The fly-

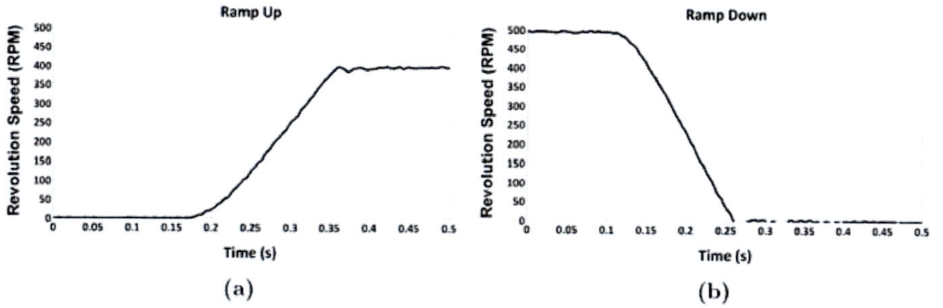

Fig. 7.9 An exemplary ramp-up of the used clutch-brake from 0 to 400 rpm is shown in (**a**). An exemplary ramp-down of the clutch-brake from 500 rpm to 0 rpm is depicted in (**b**). Figures from [1]

wheel is connected to one part of the clutch-brake that rotates accordingly. To transmit the kinetic energy, steel plates covered in special Kevlar brakepads are shifted within the clutch-brake (in the case described in this work the shift is performed pneumatically). The steel plate connected to the fly-wheel sets the second steel plate into a rotating motion as soon as the plates are forced onto each other and the brakepads connect their motion through friction. The second second steel plate is connected to a shaft that can, for example, drive a press. After the energy is transmitted, the steel plates are decoupled pneumatically and one motion cycle of the clutch-brake is finished.

Clutch-brakes exist in several variations, each with a different set of characterstics. The diameters commonly range from 40 to 100 cm, but can be even larger for special applications. Depending on the particular requirements, a clutch-brake may rotate at up to 3000 rpm with a tangential speed of up to 25 m/s. If necessary, high-tech models can perform a full braking operation within a turning angle of 20°. From these numbers it is quite obvious that the brakepads have to withstand extreme strains and are subject to significant wear.

In order to avoid a complete wear of the brakepads and the catastrophic event of the two steel plates coming into direct contact, the brake pads are replaced at fixed intervals. A condition-monitoring system (CMS) that monitors both the temperature and the thickness of the brakepads could be used to use the brakepads for longer periods of time without the risk of damaging the machine. In general, the downtimes of the machine and overall maintenance costs could be reduced. Since part of the machine is always rotating and a power line cannot easily be routed to a condition-monitoring system, an inductive energy harvesting device was investigated as an energy source.

Figure 7.9 shows exemplary ramp-up and ramp-down graphs for the clutch-brake used in these experiments. The machine can rotate at maximum speeds of up to 1000 rpm and as can be seen in Fig. 7.9 the clutch-brake action lasts less than 200 ms for this device and speeds around 400 to 500 rpm.

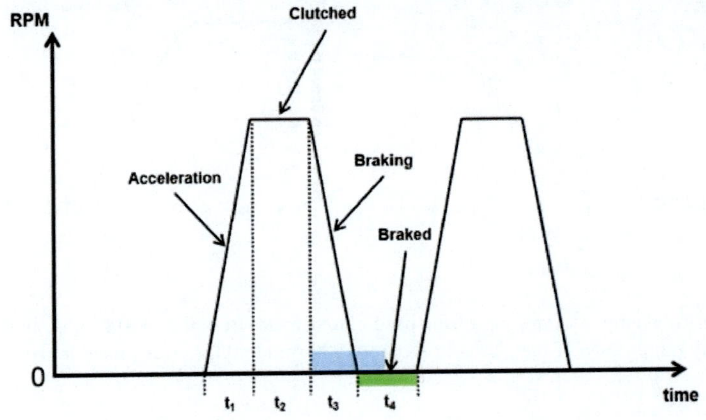

Fig. 7.10 Schematic graph showing the operating mode of the clutch-brake and a potential CMS. The time slot t_1 is the acceleration phase, t_2 is the clutched state, t_3 is the deceleration phase and t_4 is the standstill phase. The blue and green phases respectively mark the temperature and distance measurements. Figure from [1]

In the experiments, the clutch-brake was operated at 250 rpm. The operating mode of the clutch-brake and a potential condition-monitoring system is shown schematically in Fig. 7.10. The time slot t_1 is the acceleration phase, t_2 is the clutched state with a duration of 3 s in the experiments, t_3 is the deceleration phase and t_4 is the standstill phase of again 3 s. The cycle time is 6 s. The contemplated CMS would contain both temperature measurements and distance measurements on the brakepads. The blue phase in time slots t_3 and t_4 in Fig. 7.10 marks the temperature measurement with a measurement time of 2 s during the braking of the clutch-brake. The green phase in time slot t_4 in Fig. 7.10 marks the distance measurement during the standstill phase for a measurement time of 500 ms. For both measurements a sampling rate of 2 ms is assumed to provide good data.

Figure 7.11 shows the placement of the harvester (red box) on a cluth-brake. A counter-weight equal to the harvester weight needs to be placed on the opposite side of the steel plate for balance. An exemplary set of brakepads is also shown in the figure (the machines were supplied by Goizper [2]).

The harvester device is shown in detail in Fig. 7.12. In the CAD rendering you can see the magnet slider that forms a closed magnetic circuit around the coils corresponding to HAC5. The housing is design to provide space for a power management circuit and for a sensor readout circuit. In the experiments presented herein, the system only contained the power management circuit which consists of a 470 mF supercapacitor, a simple full-bridge rectifier and a hysteresis setup that turns on the regulated 3.3 V output once the capacitor is charged to 2.85 V.

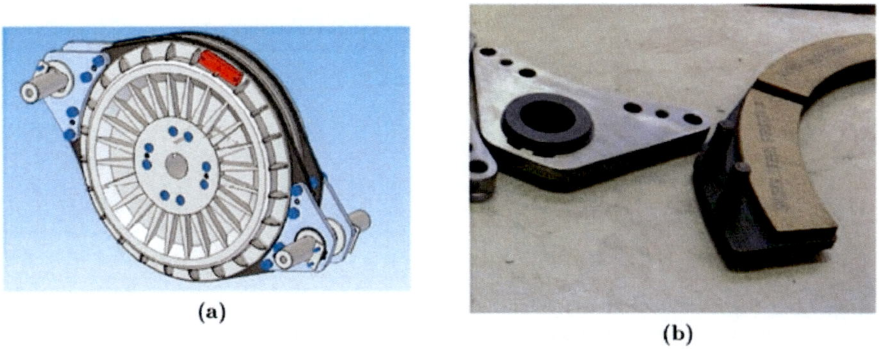

Fig. 7.11 Schematic of the used clutch-brake showing the mounting position of the harvester (red box) on the perimeter of the rotating steel plate in (a). Exemplary brakepads are shown in (b)

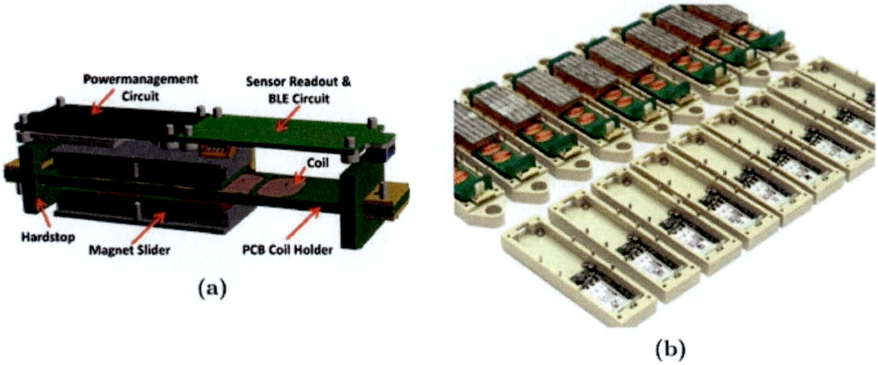

Fig. 7.12 The schematic of the inductive harvester based on HAC5 is shown in (a). Picture of several fabricated harvesting devices with a power management circuit placed in the lid in (b)

7.4 Clutch-Brake—Experimental Results

The first experiment is carried out in the laboratory by using the pneumatic piston from the beginning of this chapter as an excitation source. The piston is operated at the maximum available pressure of 7.5 bar which translates into a maximum speed of roughly 2 m/s. One motion cycle of the piston takes approximately 1.2 s. The power management circuit is designed to turn on the regulated output voltage of 3.3 V as soon as the supercapacitor is charged up to 2.85 V. Then the regulated output can supply the actual application. Once the suparcapacitor voltage drops below 2.2 V, the regulated output is switched off again. In Fig. 7.13 both the harvester mounted on the piston as well as the measured voltage at the supercapacitor (black line) and at the regulated output (red line) are shown.

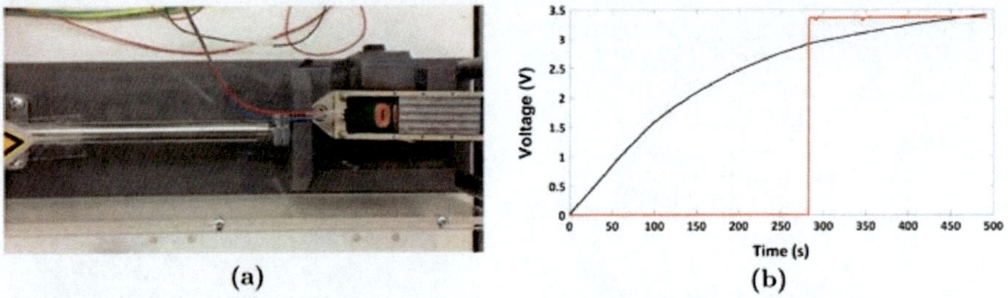

Fig. 7.13 The laboratory measurement where the harvester is attached to a pneumatic piston is shown in (**a**). The charging of the supercapacitor is shown in (**b**). The supercapacitor voltage is the black line and the regulated voltage output of the power management circuit is the red line

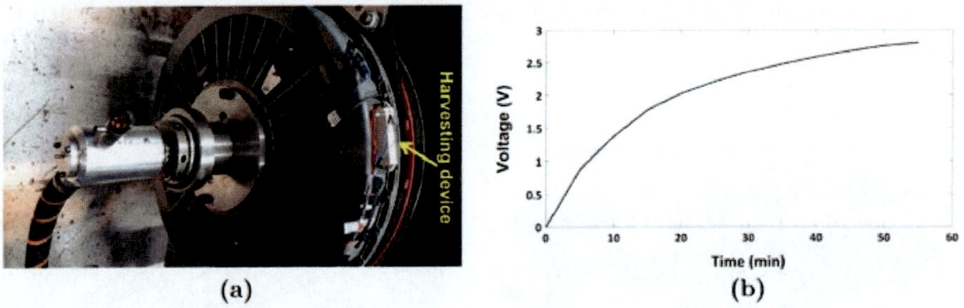

Fig. 7.14 The actual clutch-brake measurement with the attached harvester is shown in (**a**). The charging of the supercapacitor is shown in (**b**)

This was a cold-start experiment, i.e. the supercapacitor was discharged completely (< 50 mV) before being charged up again. This simulates an extended period of standstill where the capacitor would have self-discharged. The charging period under the above conditions takes 282.5 s to reach 2.85 V, which amounts to an electrical energy of 1.98 J that was stored in the supercapacitor in that time. This in turn translates into an average power of 7 mW charging the supercapacitor. In Fig. 7.13 the switch-on point of the regulated output of the circuit can be seen clearly.

The actual experiment with the harvester mounted to a clutch-brake as well as the supercapacitor voltage during the experiment are shown in Fig. 7.14. As was mentioned before, the machine was operated at 250 rpm and 3 s per phase, i.e. a cycle time of 6 s. Again a cold-start experiment is performed to charge the supecapacitor from roughly 0 V up to 2.85 V. In this experiment it takes approximately 55 min until the regulated voltage output is activated. Based on the stored energy of 1.98 J, that translates into an average power of 0.6 mW charging the supercapacitor. The warm-start charging procedure from 2.2 V up to 2.85 V takes significantly less time ($\sim$30 min).

7.4.1 Discussion

The condition monitoring system is assumed to be activated once per hour with an energy requirement of 186 mJ for the temperature and distance measurements and roughly 1600 mJ for the sleep mode operation of a microprocessor (based on measurements with a Cypress PSoC 4 BLE microprocessor) for the remainder of the hour. The amount of energy stored in the supercapacitor 1.98 J in 55 min. The harvesting device can therefore fully supply the condition monitoring system and perform a measurement every hour.

An interestin observation is that the harvester was able to charge the supercapacitor at an average power of 7 mW in the laboratory experiment, but only 0.6 mW in the field test. This difference is attributed to the large centripetal acceleration that acts on the moving magnetic stack of the harvester during the rotation of the clutch-brake. Consequently, the normal force acting on the magnet stack is increased during rotation, which leads to increased friction and limits the magnet motion.

This effect also occurs in the human gait, albeit at a much smaller scale. The foot motion during leg swing is assumed to be linear in this book, but obviously due to the pendulum motion of the leg during walking, additional accelerations perpendicular to the magnet motion within the harvester occur, that can have an impact on friction. With additional acceleration measurements, this effect can be incorporated into the system model presented in Chap. 2 in future work and it can be analysed, how the power generation capability of the harvester is affected.

References

1. D. Hoffmann, K. Ylli, A. Willmann, D. Stojakov, Y. Manoli, M. Tijero, M. Mondragon, Energy harvesting devices for condition monitoring applications of pneumatic combined clutch-brakes. J. Phys.: Conf. Ser. **1052**, 012076 (2018)
2. Goizper, Goizper clutch-brakes (2017), http://www.goizper.com/de/industrie/bremsen-und-kupplungen/pneumatisch. Accessed 03 March 2017
3. K. Ylli, Energy harvesting from the swing motion of the human leg (2019)

Chapter 8
Conclusion and Outlook

8.1 Conclusion

The major part of this work deals with the development of a system model, which is based on a differential equation of motion (eom). The eom contains all forces that act on the moving magnet stacks and is implemented in the numerical simulation environment Matlab/Simulink in order to determine the motion of the magnets. From the relative motion between the magnets and the coils of the device, the induced voltage and subsequently the power dissipated in an ohmic load can be calculated. Through this simulation setup, the expected power output of a given device can be predicted before fabrication.

Four different harvester architectures, i.e. different magnet-coil-arrangements, are investigated in this work. Two architectures consist of cylindrical magnet and coil structures, while the remaining two are rectangular. The device volume including housing is defined as 77 mm in length, 41.5 mm in width and 15.75 mm in height. The active transducer volume without the housing is a little less: 71 mm x 12.5 mm x 37.5 mm. An optimization of the geometrical parameters was performed for each architecture in order to find the parameter set that gives the largest average power output. The largest power output was predicted for the rectangular architecture with a closed magnetic circuit HAC5.

Following the optimization, a first generation of devices was fabricated as closely to the optimized geometries as possible and characterized experimentally on a treadmill. The voltage was measured across an ohmic load equal to the coil resistance of the respective harvester. Measurements were performed for four different treadmill speeds (4, 6, 8 and 10 km/h) and two test persons. At the normal walking speed of 4 km/h and for test person "1", an average of 4.64 mW was obtained for HAC1, 5.62 mW for HAC1 XL (an additional version of HAC1 with longer coils), 0.39 mW for HAC3, 4.89 mW for HAC5 and 1.91 mW for HAC6. The largest experimental average power output of 10.32 mW was measured for HAC1 XL at a motion speed of 10 km/h.

K. Ylli and Y. Manoli, *Energy Harvesting for Wearable Sensor Systems*,
Springer Series in Advanced Microelectronics 62,
https://doi.org/10.1007/978-981-33-4448-8_8

The largest power densities for the first generation experimental devices were also achieved by HAC1 XL. Considering only the active volume, i.e. the volume occupied by the magnets, coils and the channel through which the magnets move, a power density of 627.2 μW/cm^3 was achieved at 4 km/h and a power density of 1027.9 μW/cm^3 was achieved at 6 km/h. Under the same conditions, the best rectangular architecture HAC5 achieved power densities of 142.9 μW/cm^3 at 4 km/h and 235.5 μW/cm^3 at 6 km/h. For the second test person "2", the power outputs were generally lower. This can likely be explained by a longer step size and the resulting smoother motions required to keep up a given speed as opposed to test person "1" with a smaller step size.

The investigated devices show different behaviors as far as the test persons are concerned. It is obvious that the power output of a device will vary with the test person and the corresponding gait pattern. The variations however, show a certain pattern. For those devices, where the magnet mass is comparatively small, e.g. HAC1 and HAC3, the power output for test person "1" is larger than for "2" throughout all measurements.

In some particular cases, the power output is almost twice that of "2". In the case of the architectures with heavier magnet stacks, e.g. HAC5 and HAC6, the difference in power output between the two test persons is relatively small. For the measurements performed at 8 km/h in fact, the order is reversed and test person "2" generates a larger power output than "1".

This observation indicates the importance of the inertial mass in this kind of system. With a larger inertial mass, the difference in the gait pattern appears to lose influence. The kinetic energy is proportional to the moving mass and it follows that the damping of the motion, both mechanical and electrical, has a reduced relative effect on magnet motion as the mass increases.

Based on the experimental data from the first generation of devices, a number of bugfixes and model revisions were performed. As was shown in Chap. 4, the deviation between simulations and experiments was reduced greatly, to values below 20%, in many cases below 10%. On that basis a new set of optimizations was performed in Chap. 5, as it could be assumed, that the first generation designs were not optimal before the model revision.

The second optimization run showed that for the cylindrical HACs, the first generation was quite close to the optimum (HAC1 XL is only 24% off the newly calculated design in terms of power output). However, for the rectangular setups it promised great improvements. Based on the optimization run and the fact that the assembly process of the cylindrical harvesters is very difficult due to the opposing magnet orientation of the transducer, only the rectangular HACs were considered for a second generation. The newly fabricated and assembled devices were characterized on a treadmill following the exact same procedure as was the case for the 1st generation.

The new devices of HAC5 and HAC6 showed a tremendous improvement. HAC5 outperformed its first generation design by 432%, achieving an average power output of up to 26 mW at a motion speed of 4 km/h for test person "1". At 6 km/h the device reached its peak average power output of 43 mW for test person "1". HAC6 showed an even greater improvement over its first generation design at 946% which translates

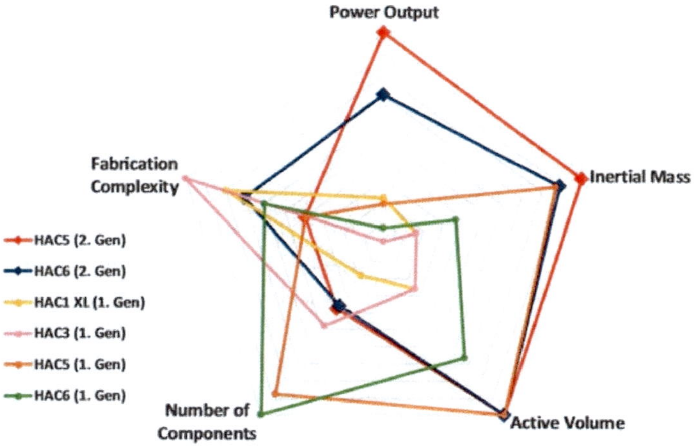

Fig. 8.1 Net diagram comparing the fabricated devices from Chaps. 4 and 6. The values are normalized to the maximum in each category. The power output is the only parameter that is required to be large. In the case of the remaining parameters, a smaller value is advantageous

to an average power output of nearly 20 mW at at a motion speed of 4 km/h for test person "1". This device reached its peak average power output of 31.7 mW at a motion speed of 10 km/h for test person "1". The difference in the walking pattern of the two test persons is difficult to link to the harvester behavior for the new devices. The power output seems to be quite close for the two persons in some measurements, but rather different in other measurements, without any noticeable pattern. As was explained in the discussion of Chap. 6, some mechanical issues appeared during testing of the 2nd generation harvesters, which appear to have affected the power output of some measurements and may have disturbed the power output patterns. An increased data set with a larger number of test persons is required for a conclusive analysis in the future.

The average power density calculated from the experimental results has improved analogously. Considering the active volume of 34.22 cm^3, the 2nd generation HAC5 reaches a maximum average power density of 1260 $\mu W/cm^3$ at a motion speed of 6 km/h for test person "1". With the same active volume, the 2nd generation HAC6 reaches its maximum average power density of 930 $\mu W/cm^3$ at a motion speed of 10 km/h for test person "1".

When considering the total volume (including the housing) of 50.33 cm^3, the 2nd generation HAC5 reaches a maximum average power density of 860 $\mu W/cm^3$ at a motion speed of 6 km/h for test person "1". The 2nd generation HAC6 reaches its maximum average power density of 630 $\mu W/cm^3$ at a motion speed of 10 km/h for test person "1".

All power figures have shown a great improvement over the first generation of HAC devices. A graphical comparison of the first and second generation devices is given in Fig. 8.1. Five categories are used to compare the devices and the values are normalized to the maximum in each category. For the power output, data measured at 4 km/h for test person “1” was used. For the first generation HAC1 XL and HAC5 perform the best in this category. The second generation devices improved greatly upon the first generation in terms of power output. The 2nd generation HAC5 outperformed HAC6 noticeably, although HAC6 showed a nearly tenfold increase in power output. While still within the size constraints, the rectangular architectures require the most space in the shoe sole. They also contain a comparatively large number of components (magnets, flux guides and coils). The mass of the harvesting devices is also the largest for the rectangular architectures which can be regarded as a disadvantage considering the user comfort. In terms of fabrication complexity, the cylindrical architectures rate very high. This is explained by the difficulties in the magnet gluing process which has to withstand the opposing forces from the magnets and caused several device failures during testing. HAC5 is comparatively simple to fabricate as the magnets can easily be placed on the back iron sheet due to magnetic attraction. Only the combination of the upper and lower part of the magnetic circuit requires particular attention. Overall, HAC5 performs the best in this comparison.

A comparison of different devices from literature with the first generation HACs presented in this book is given in Fig. 8.2a. The fabrication complexity is based on the subjective experience obtained during the course of this work. One example is HAC1 which contains only a small number of components, but still is quite difficult to fabricate because magnets have to be glued together against strong opposing forces. Other devices, where the magnets hold in place, e.g. due to attractive forces with a piece of flux guiding material, are less complex and thus assigned a smaller number. The remaining parameters: power output, active volume, number of components and inertial mass are taken from the corresponding papers or estimated, where a concrete number was not given. Data at 6 km/h was used for Fig. 8.2 for all devices except Shen et al. [1] which did not give the exact walking speed.

It can be seen in Fig. 8.2a that the device from Carroll et al. [2] performs approximately as well as the presented HACs in terms of power output. In fact, they were able to generate a larger power output than HAC1 when not connecting the coils in series and looking at them separately, with separate loads etc. This comparison however, only contains setups where the coils are connected in series, so that they require only one rectifying circuit. HAC5 performs well in terms of power output and fabrication complexity, but has a heavy inertial mass and requires a comparatively large number of magnets and coils as indicated in Fig. 8.2a.

When looking at the average power outputs for the second generation HAC5 and HAC6 in comparison to devices from literature, it is apparent, that HAC5 outperforms all other devices by a large margin, fortifying the good results presented in this book.

In terms of fabrication complexity and inertial masses, little changes in comparison to the first generation devices. The 2nd generation HAC5 requires fewer components than before. However, due to the thinner back iron the fabrication complexity is slightly larger because of the difficult placement of the braces that hold the two sides of the magnetic circuit apart. The second generation HAC6 has increased in volume and inertial mass over the first generation design as it makes full use of the available width within the housing due to the custom fabricated magnets and

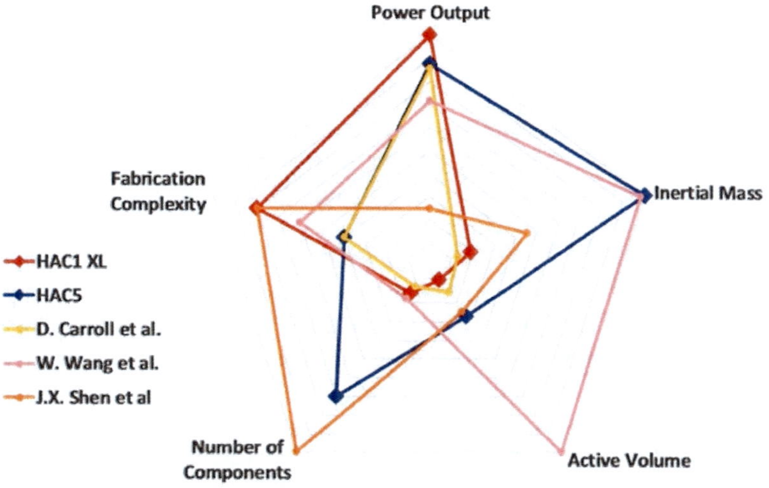

(a) 1st generation HACs vs. Literature

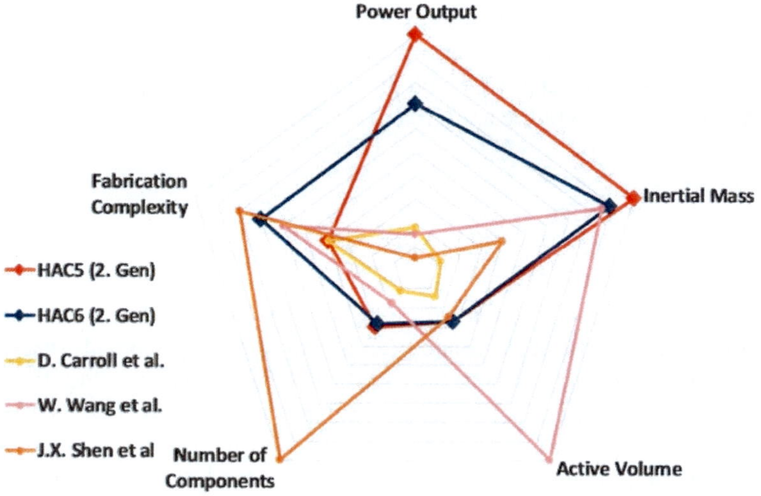

(b) 2nd generation HACs vs. Literature

Fig. 8.2 Net diagram comparing the first generation devices from Chap. 4 with devices from literature in (**a**) and comparing the second generation HACs from Chap. 6 to devices from literature in (**b**). The values are normalized to the maximum in each category. The power output is the only parameter that is required to be large. In the case of the remaining parameters, a smaller value is advantageous. "D. Carroll et al." refers to [2], "W. Wang et al." refers to [4] and "J. X. Shen et al." refers to [1] coils (unlike the first generation design which used readily available magnets with a smaller width).

In terms of power densities, device [2] reaches a value of 447 $\mu W/cm^3$ at a motion speed of 6 km/h (considering only the active volume without a housing). The second generation HAC5 reaches nearly three times that at 1260 $\mu W/cm^3$, resulting in another benchmark performance for flat inductive harvesters for integration in the shoe sole.

Deviating from the aim to power electronic systems using human motion, a range of laboratory experiments were performed on a pneumatic piston in Chap. 7 using the first generation harvesters. They proved, that the system model developed in this work is not suitable for acceleration data that contains large and narrow peaks of only a few milliseconds. The model lags behind in its reaction to the input data and thus the simulated power output does not match the experimentally obtained data. The experiments did prove though, that such linear harvesting devices can be used for a wide range of industrial applications in addition to body-worn systems. Particularly at high accelerations, the power output can grow noticeably beyond what is possible when the human gait is the energy source. The largest power output using the first generation HACs on the piston was achieved with HAC5 using an ohmic load equal to the coil resistance (480 Ω) at an average of 13.35 mW.

Two industrial applications, pneumatic pistons and clutch-brakes were presented in Chap. 7. A suitable application on the human body and opportunities to improve upon this work in the future are given in the following sections.

8.2 Viable Human Body Application

The aim of the human motion harvesting devices fabricated in this work is to be able to power body-worn electronic systems autonomously and without the need for battery replacement or maintenance. While the generated power does not suffice to charge a powerful device like a smartphone, it may very well be used for a low-power system.

An exemplary body-worn system is the fitness-tracker, which counts the user's steps and speed. Other systems may gather biological data from the user, e.g. for medical monitoring in cases of illness or recovery. As long as a high data rate is not required, a sensor system can transmit the data wirelessly using low-power standards like "Bluetooth low-energy" (BLE). Building on these descriptions, a biological data tracker is being developed at Hahn-Schickard, which could be integrated into the shoe. Using light-emitting diodes (LED) directly placed on the skin, light is emitted into the body. The blood pulse causes a change in the reflection of the light which can be detected using photo-diodes. This type of measurement is not only possible at the finger tips, where the blood vessels run close to the skin surface (this measurement type is generally used in hospitals). At the ankle blood vessels also run close enough to the skin, for a photo-diode to detect a pulse and in this setup, the biological tracker is placed near the shoe, where energy can be generated using the leg swing. This

Table 8.1 Key components and properties for a biological tracker that can measure the blood pulse at the ankle. Table from [3]

Component	Manufacturer	Properties
CYBLE-0144008-00	Cypress	Arm Cortex M0 MCU with Bluetooth low-energy, 1.3 μA standby, 17 mA Transmit/Receive current
MAX30100	Maxim Integrated Products	Integrated LED Driver for infrared (IR) and red LED with selectable output current (0–50 mA), a photodiode and an integrated precision measurement circuit (0.6–1.2 mA)
SFH7060	OSRAM Opto Semiconductors	Integrated LED chip with green (25 mA), red (50 mA), IR (60 mA) LEDs and a photodiode. Separate measurement circuit is required

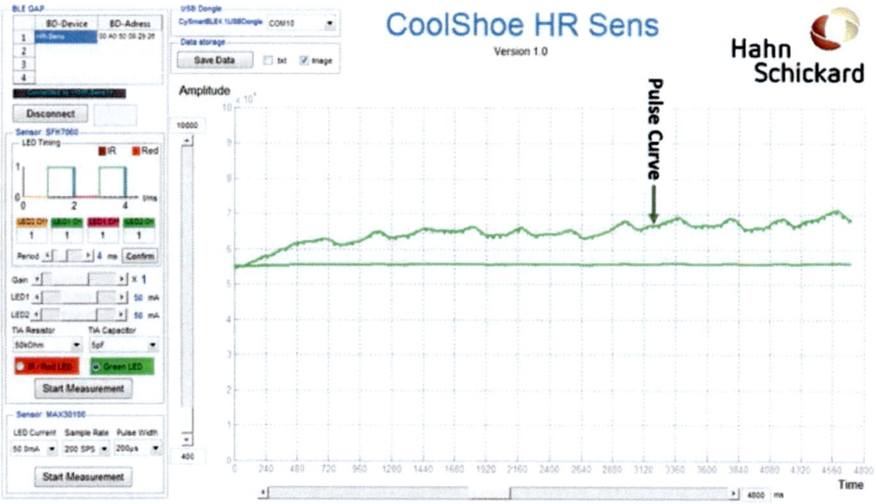

Fig. 8.3 Screenshot of the pulse measurement software implemented in Matlab by Hahn-Schickard. The green line shows the unfiltered sensor signal from exemplary measurements at the forehead. The single impulses indicating the pulse can be noticed. Figure from [3]

blood pulse measurement system can be integrated on a flexible circuit board and can be fastened to the ankle by a Velcro-strap which is attached to the shoe. This guarantees the power supply and the data can be transmitted to a handheld-device.

Key components of this biological tracker are listed in Table 8.1. A pulse was measured at the ankle. The sensors can also measure blood oxygen levels, as is done by devices used in hospitals, which are placed on the fingertip. However, it has not yet been confirmed, that these sensors can detect the blood oxygen levels reliably at the ankle. The unfiltered sensor information as obtained by Hahn-Schickard during exemplary measurements on the forehead is shown in Fig. 8.3.

The signal peaks caused by the blood pulse can be seen in the marked curve. In a filtered signal, the blood pulse would be detected automatically. However, the device is not yet optimized for low power consumption and thus cannot be operated autonomously. By using the sleep modi of the micro-controller and the RF chip, and setting up the blood pulse sensor correctly, it is expected to run the system at a continuous power consumption of a few milli-Watts, if a data transmission is not required several times per second.

While human motion was the energy source of choice in this book, other viable energy sources exist in industrial applications, wherever there are linear motions. The pneumatic piston and clutch-brake from Chap. 7 are only two such examples.

8.3 Outlook

Building on this work, it is possible to proceed in various directions. One possible approach is to consider new harvester architectures. As an example, a magnet-in-coil architecture was simulated, where the magnets and the coils around them are not cylindrical, but rectangular. This allows a more efficient use of the space within the housing as there is no empty volume due to cylindrical coils in a rectangular housing. While this system model is not yet verified, a geometrical optimization run resulted in a predicted power output of up to 15.8 mW when using the full width of the housing as was the case for HAC5. This shows, that there are still architectures that would require further investigation.

Another task for future work is the investigation of new modeling approaches that are less sensitive to the input data and possibly more suitable for the modeling of the real-world physics of such a complex system. In this work universal Simulink blocks were used for the system model. However there are toolboxes for Simulink, e.g. "SimScape" and "SimMechanics" which are specifically designed to model mechanical systems and their motion. Other modeling environments like "Modelica" may be an alternative.

Even more important from the point of view of the authors is the fact, that until now the presented harvesters were stand-alone devices, physically separated from the application, without any other function than to generate energy. The simulations were set up accordingly, only considering ideal ohmic loads. A major point for future work is the integration of an electrical simulation in the system model using special toolboxes. This way, a device could be optimized end-to-end based on realistic power management circuits and not based solely on ohmic loads. Ohmic loads are a perfectly suitable for the optimization and comparison of devices, however, these results cannot tell you what is the actual net power that is available for your application. Figure 8.4 shows a first implementation of a power management circuit in the "SimScape" toolbox of the Matlab/Simulink environment. That circuit contains a full-bridge rectifier, a 0.2 F supercapacitor and a hysteresis function with thresholds for the regulated voltage output at 3.3 and 2.2 V, as well as a variable load with a power output of 400 mW that simulates the power consumption of an application. Once linked to the "normal" system model in Simulink, an end-to-end optimization based on this coupled model becomes reality and the optimization procedure could be

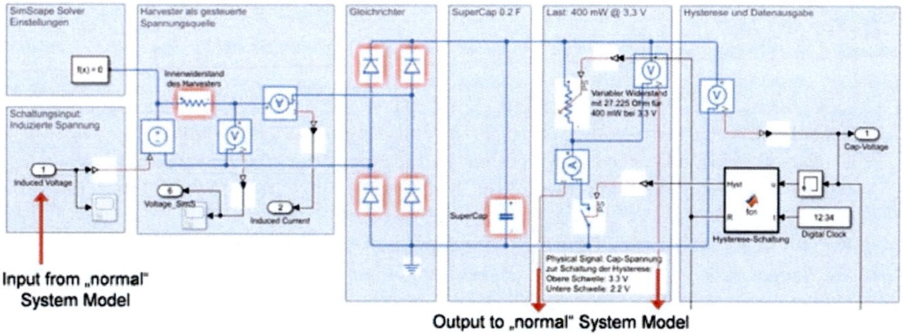

Fig. 8.4 Screenshot of a first implementation of a power management circuit using the "SimScape" toolbox. The model contains a full-bridge rectifier and hysteresis thresholds for the regulated voltage output at 3.3 and 2.2 V, as well as a variable load with a power output of 400 mW

tailored to optimize both mechanical and electrical parameters. Future devices could therefore be fabricated to match the actual needs of an application.

The cost of a fabricated harvester architecture was no priority in this work. However, in future work, the cost of the used components could be taken into account during the geometrical parameter optimization. Two of the greatest cost factors are the magnets made from rare earths and the copper used in the coils. A first multi-objective optimization was performed, where the maximization of the power output remained the first objective.

The second objective is to minimize the magnet mass. This is a simple optimization for cost. The optimization produced several results that aimed at a low magnet mass. An exemplary HAC1 design generates an average power output of 4.7 mW, approximately 64% of the largest possible power output predicted in Chap. 5, at only 36.3% of the weight (5.98 g as opposed to 16.47 g in the maximum power output design). Future geometrical optimization procedures can be adapted accordingly, to take more factors than just the maximized power output into account.

A similar approach can be chosen to reduce the overall size of the system. Depending on a given application, the harvesting device can be tailored to provide the required power output at smallest possible size. A smaller form factor results in smaller modifications to the structure of the shoe.

Finally, it is important that these devices are integrated into shoes. Previous devices of the same size have been integrated into running shoes, proving that the integration is possible without a significant loss in user comfort. Shoe soles inherently designed to house a harvester could provide a further improvement to the wearer's comfort. Once integrated, the longevity of the devices could be monitored, as to this date, there is no information on reliability. The harvesters can be coupled to a sensor system, like the above described biomedical tracker, to provide a benefit to the wearer in today's connected world.

References

1. J.-X. Shen, C.-F. Wang, P.C.-K. Luk, D.-M. Miao, D. Shi, C. Xu, A shoe-equipped linear generator for energy harvesting. IEEE Trans. Ind. Appl. **49**(2), 990–996 (2013)
2. D. Carroll, M. Duffy, Modelling, design, and testing of an electromagnetic power generator optimized for integration into shoes. Proc. Inst. Mech. Eng., Part I: J. Syst. Control. Eng. **226**(2), 256–270 (2012)
3. K. Ylli, Energy harvesting from the swing motion of the human leg (2019)
4. W. Wang, J. Cao, N. Zhang, J. Lin, W.-H. Liao, Magnetic-spring based energy harvesting from human motions: design, modeling and experiments. Energy Convers. Manag. **132**, 189–197 (2017)

Appendix

A.1 Full Experimental Dataset from Treadmill Measurements

The data described in Chap. 4 is a compact representation of a wide range of treadmill measurements. The full set of graphs generated from the experimental data could not be presented in the full text. Instead, it is shown in this section of the appendix. The following graphs show voltages and power outputs for the investigated harvester architectures. The graphs are given for both test persons and all four treadmill speeds used in the experiments (4, 6, 8 and 10 km/h). As was the case in the main body of this work, out of the total of 30 s of data obtained for each measurement, only 1.5 s are shown in the graphs in order to get a better look at the waveform (Figs. A.1, A.2, A.3, A.4, A.5, A.6, A.7, A.8, A.9, A.10 and A.11).

K. Ylli and Y. Manoli, *Energy Harvesting for Wearable Sensor Systems*,
Springer Series in Advanced Microelectronics 62,
https://doi.org/10.1007/978-981-33-4448-8

A.1.1 HAC1 for Test Person "1"

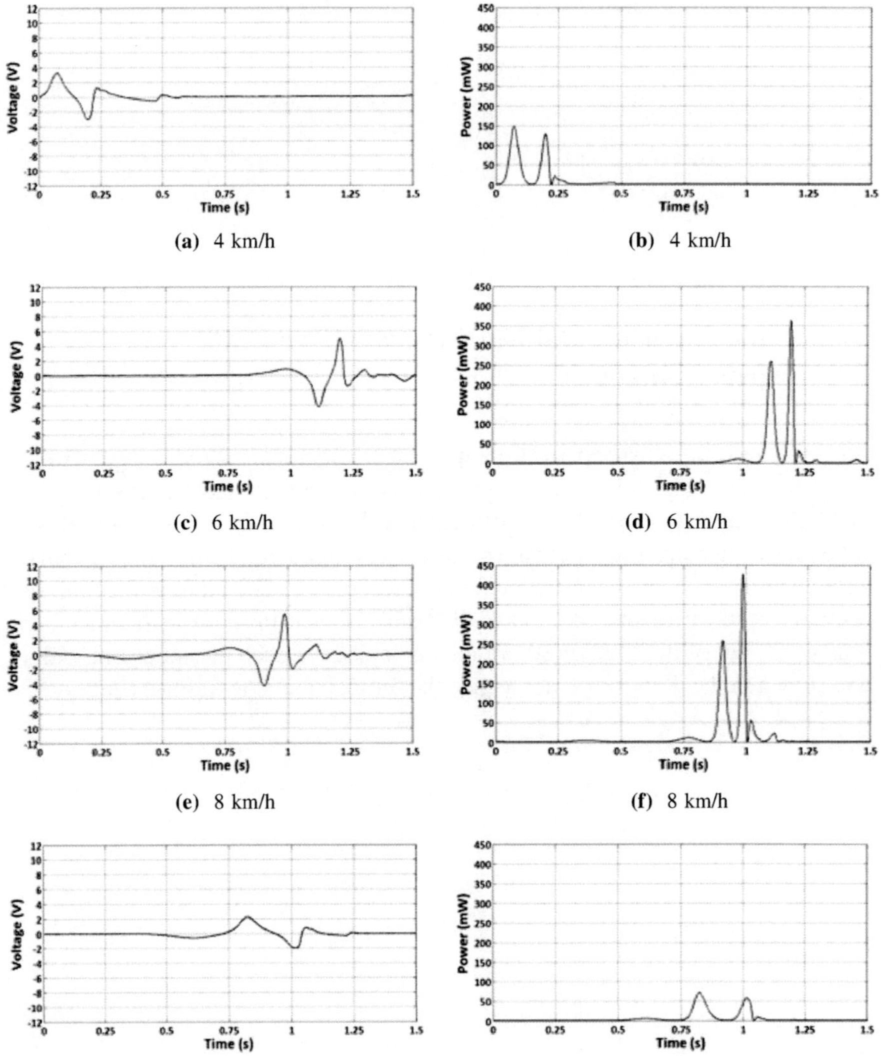

(a) 4 km/h (b) 4 km/h

(c) 6 km/h (d) 6 km/h

(e) 8 km/h (f) 8 km/h

(g) 10 km/h (h) 10 km/h

Fig. A.1 Voltage (on the left side) and power output (on the right side) data obtained experimentally for HAC1 (one magnet stack) and test person "1". The load resistance of 70 Ω is equal to the total coil resistance and identical across all measurements. Figures from [1]

A.1.2 HAC1 for Test Person "2"

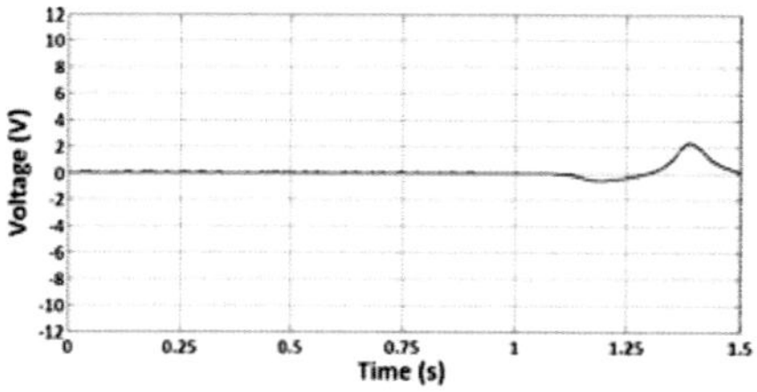

(a) 4 km/h

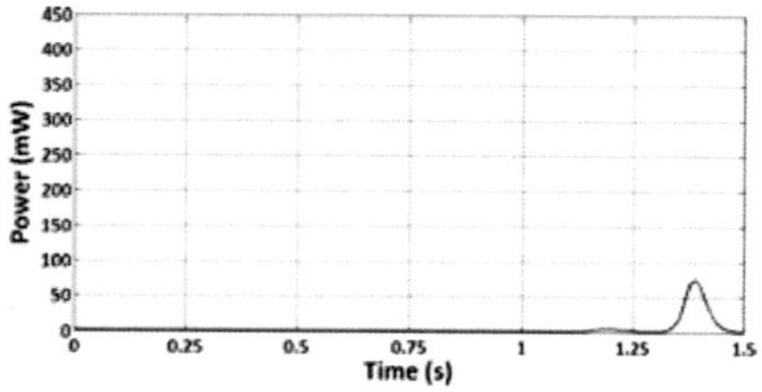

(b) 4 km/h

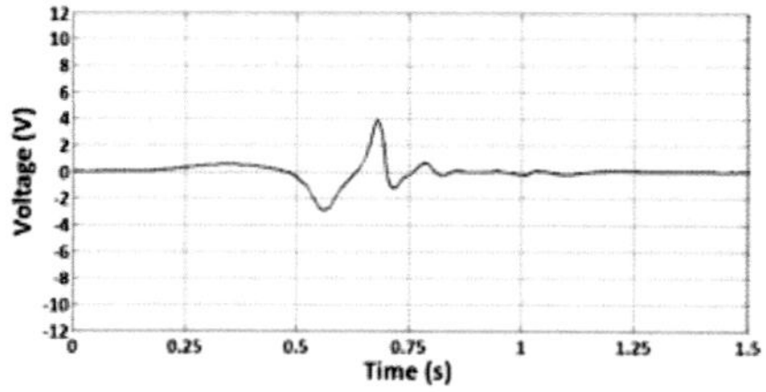

(c) 6 km/h

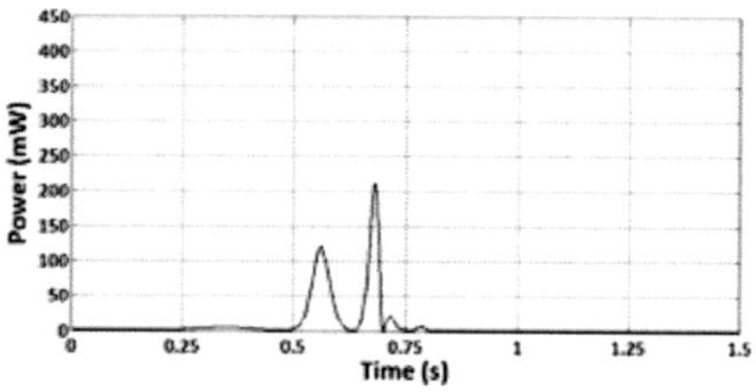

(d) 6 km/h

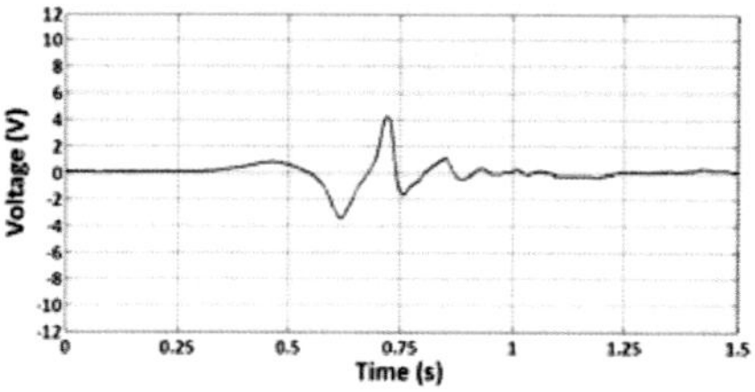

(e) 8 km/h

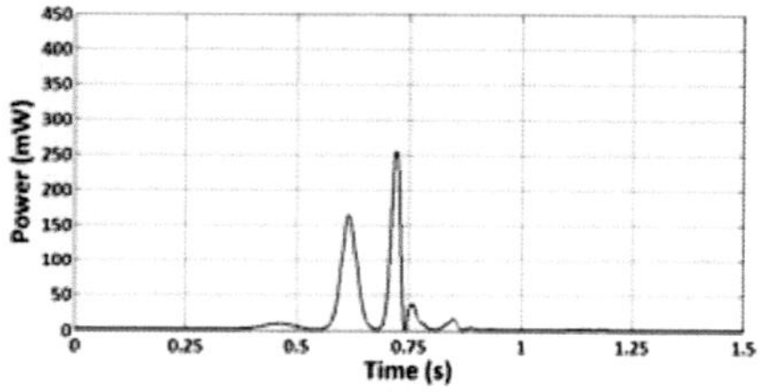

(f) 8 km/h

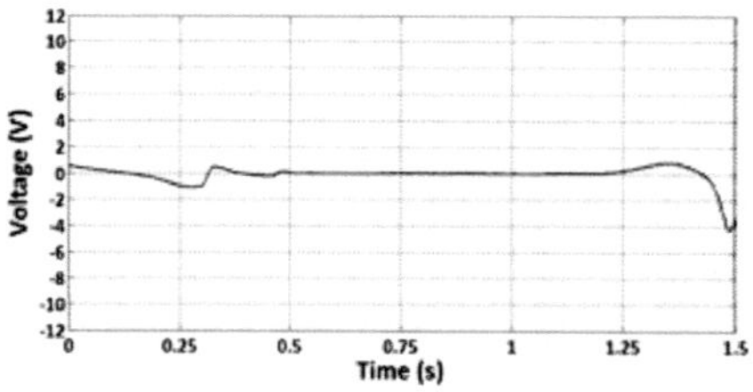

(g) 10 km/h

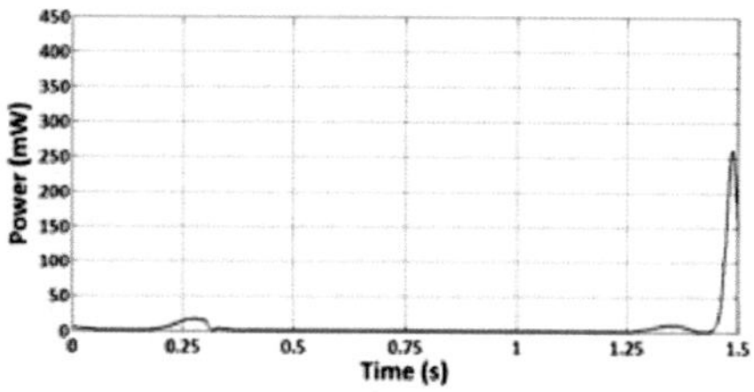

(h) 10 km/h

Fig. A.2 Voltage (on the left side) and power output (on the right side) data obtained experimentally for HAC1 (one magnet stack) and test person "2". The load resistance of 70 Ω is equal to the total coil resistance and identical across all measurements. Figures from [1]

A.1.3 HAC1 XL for Test Person "1"

For HAC1 XL, there is only experimental data for test person "1", as there were device failures during the experiments.

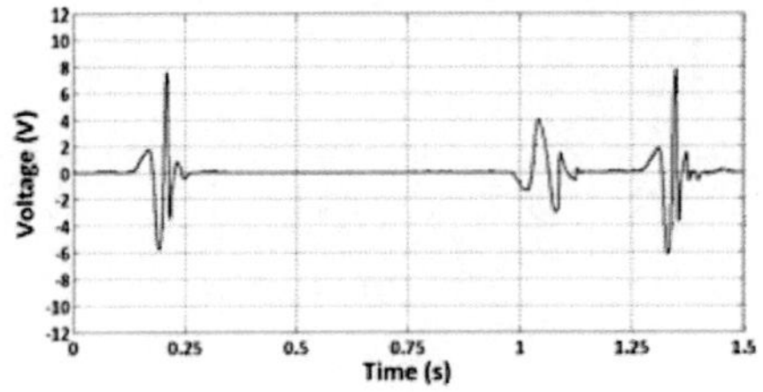

(a) 4 km/h

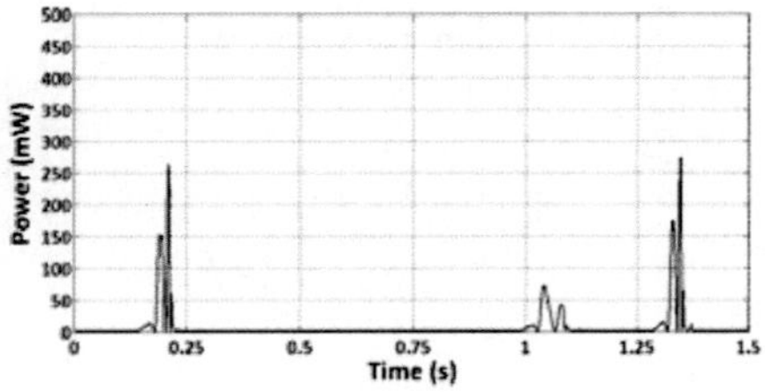

(b) 4 km/h

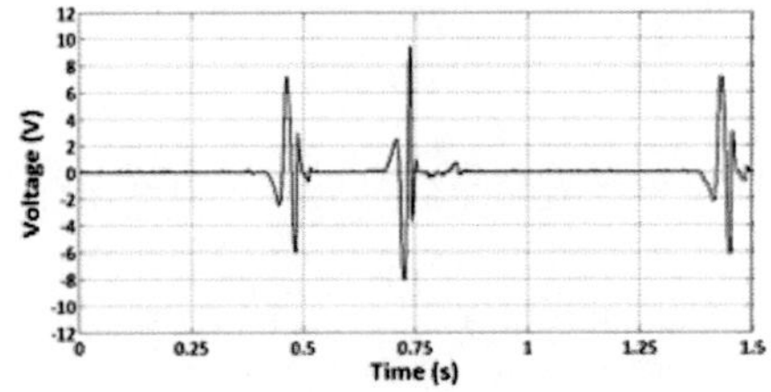

(c) 6 km/h

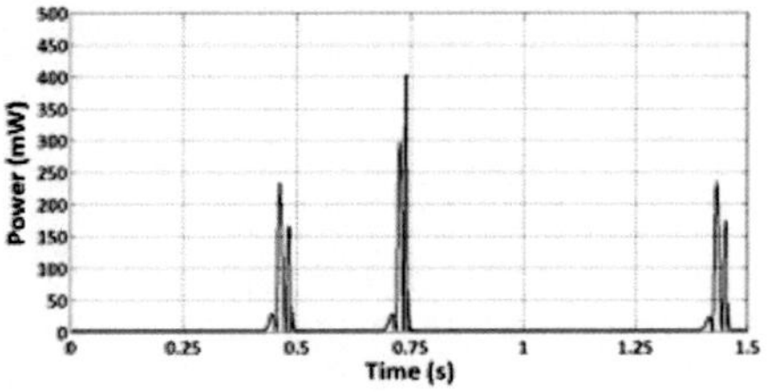

(d) 6 km/h

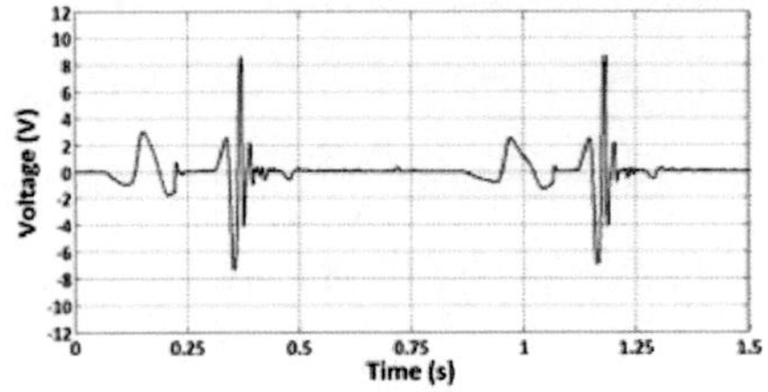

(e) 8 km/h

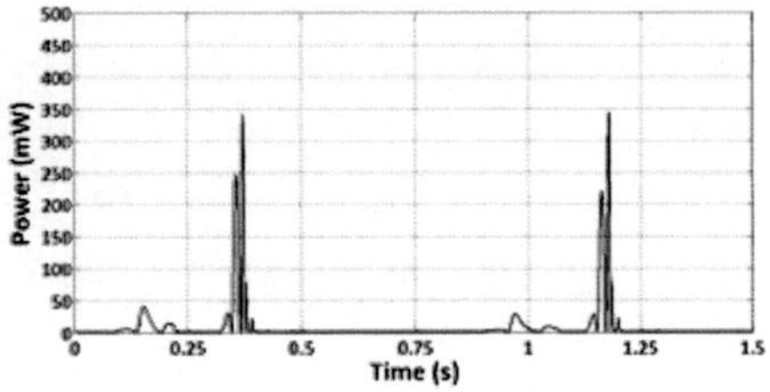

(f) 8 km/h

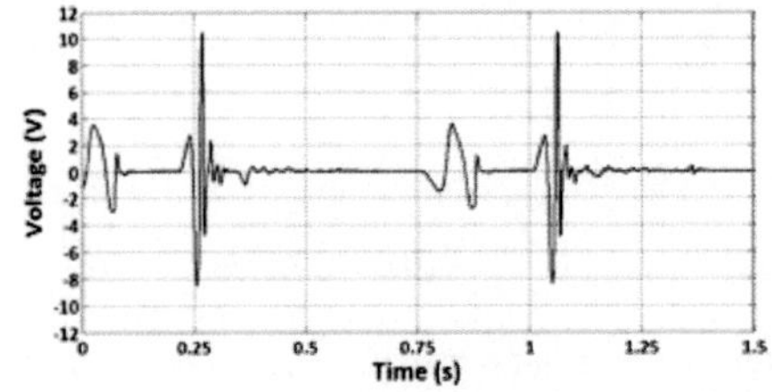

(g) 10 km/h

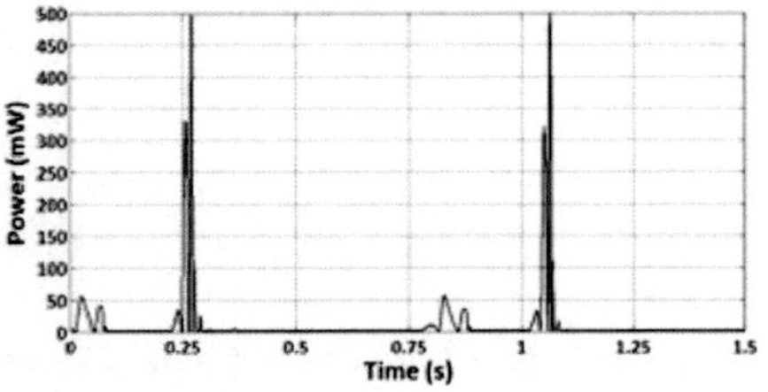

(h) 10 km/h

Fig. A.3 Voltage (on the left side) and power output (on the right side) data obtained experimentally for HAC1 XL (one magnet stack) and test person "1". The load resistance of 220 Ω is equal to the total coil resistance and identical across all measurements. Figures from [1]

A.1.4 HAC3 for Test Person "1"

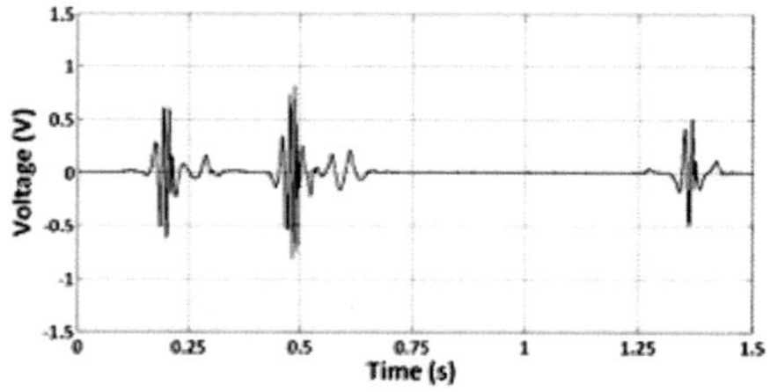

(a) 4 km/h

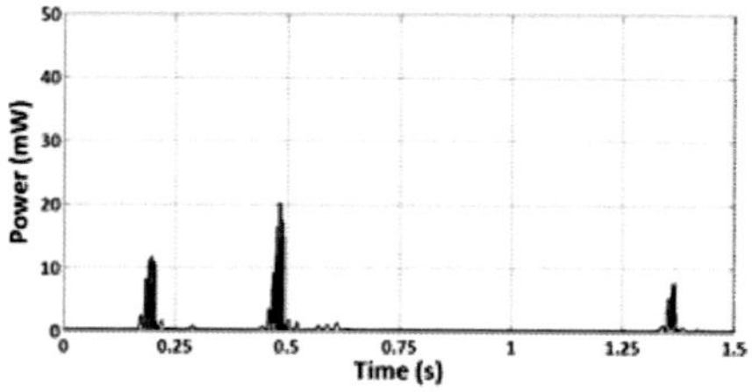

(b) 4 km/h

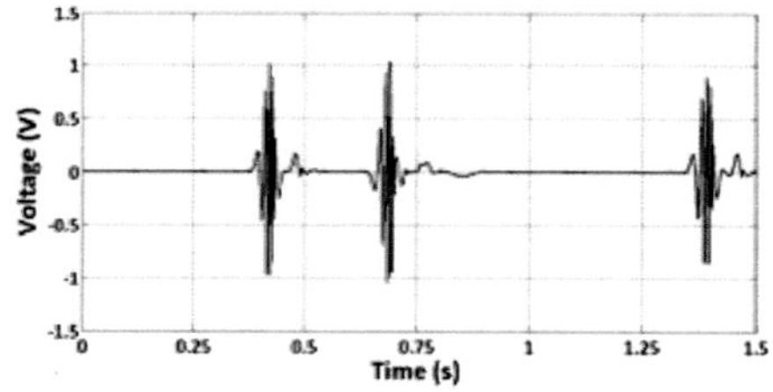

(c) 6 km/h

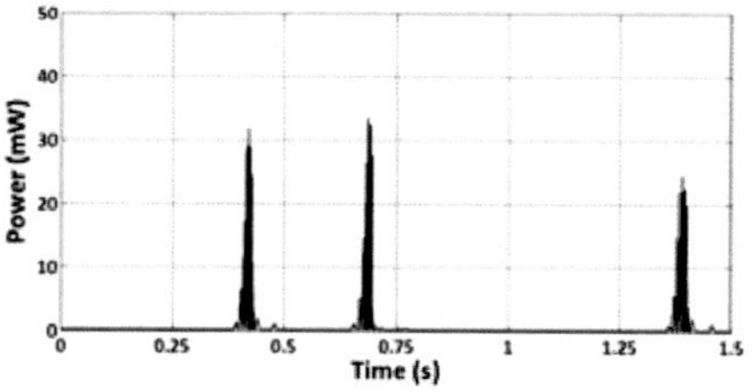

(d) 6 km/h

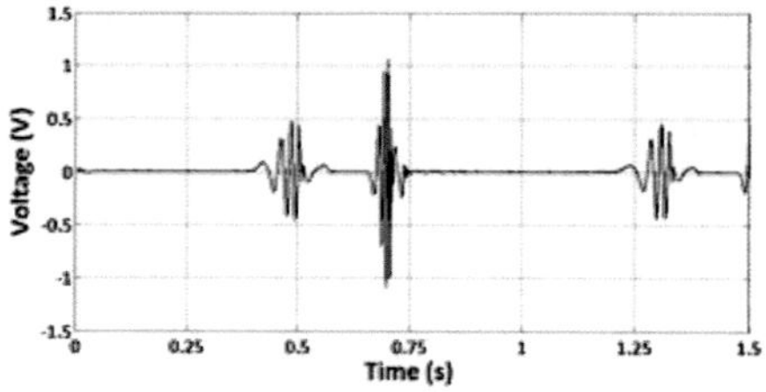

(e) 8 km/h

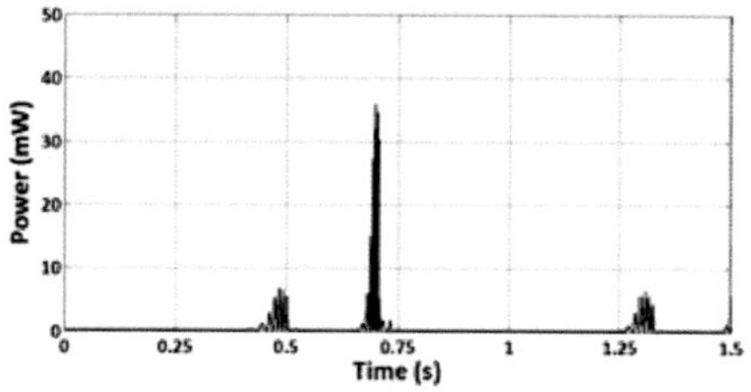

(f) 8 km/h

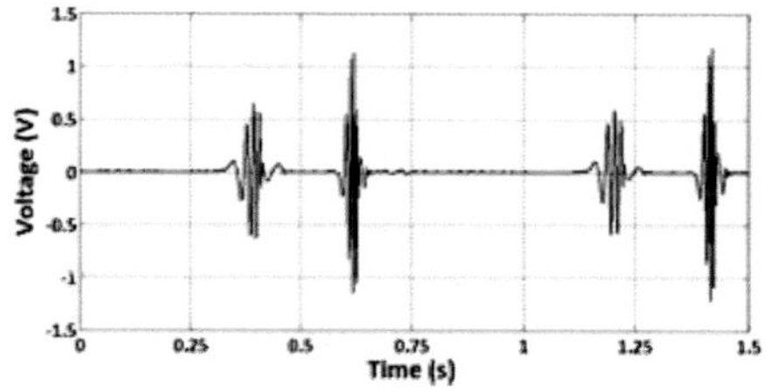

(g) 10 km/h

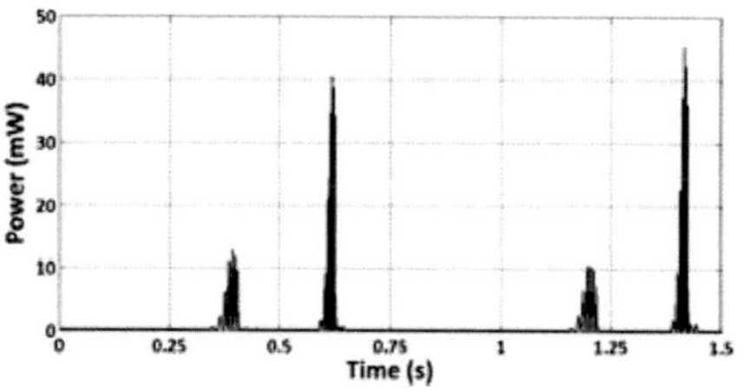

(h) 10 km/h

Fig. A.4 Voltage (on the left side) and power output (on the right side) data obtained experimentally for HAC3 (one magnet stack) and test person "1". The load resistance of 33 Ω is equal to the total coil resistance and identical across all measurements. Figures from [1]

A.1.5 HAC3 for Test Person "2"

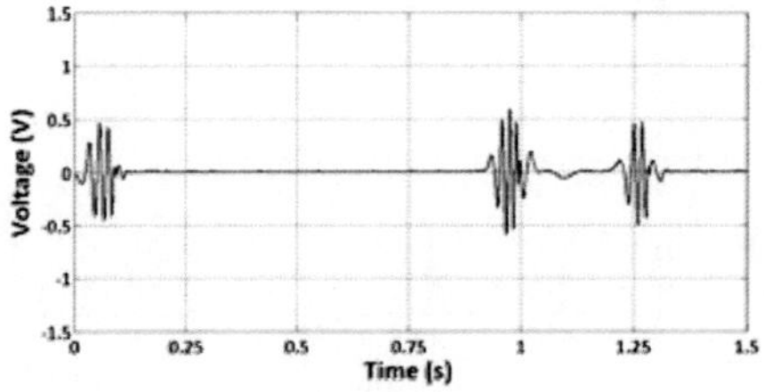

(a) 4 km/h

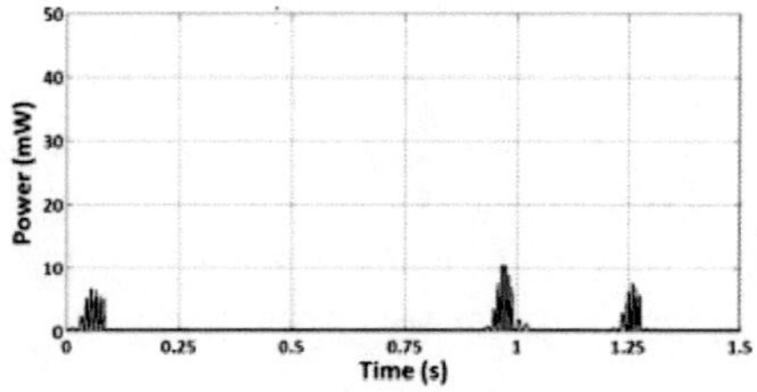

(b) 4 km/h

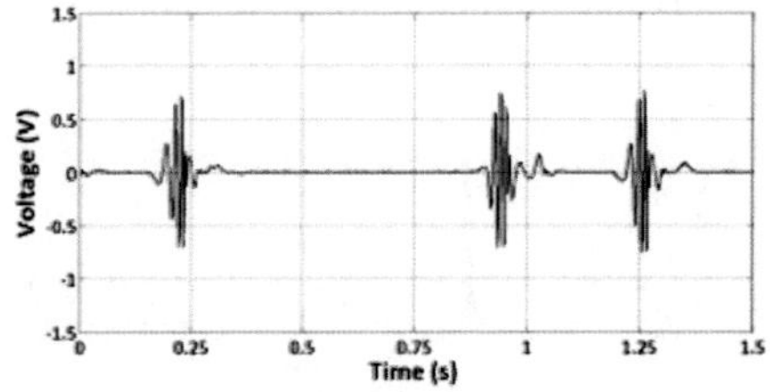

(c) 6 km/h

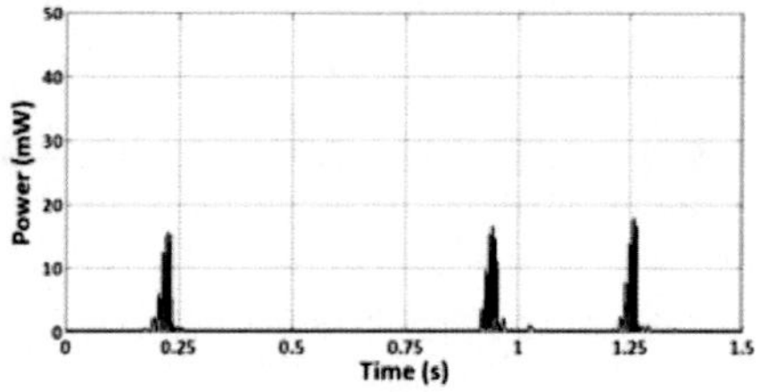

(d) 6 km/h

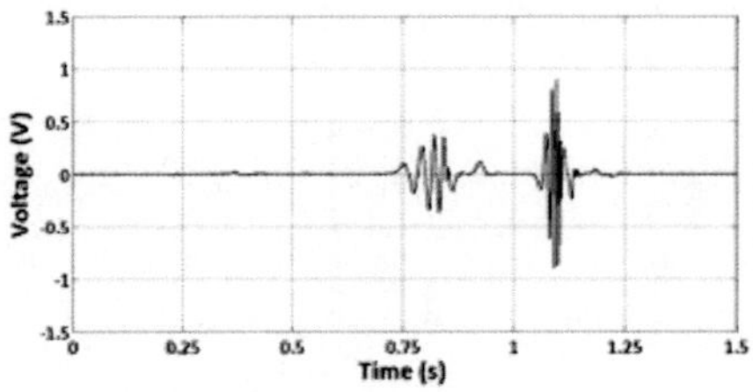

(e) 8 km/h

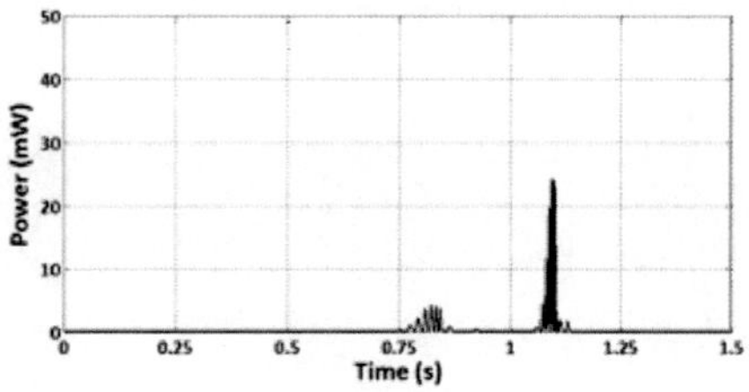

(f) 8 km/h

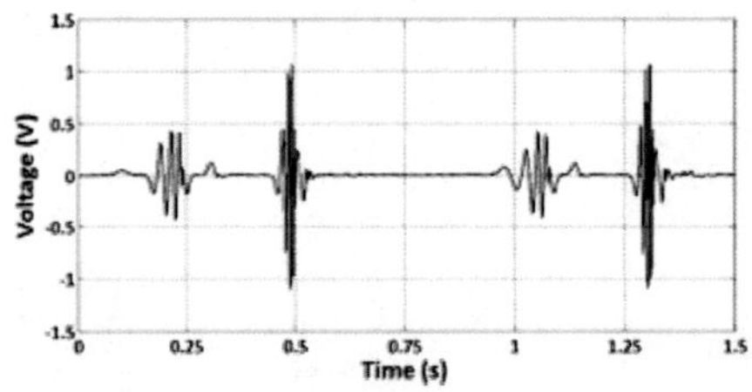

(g) 10 km/h

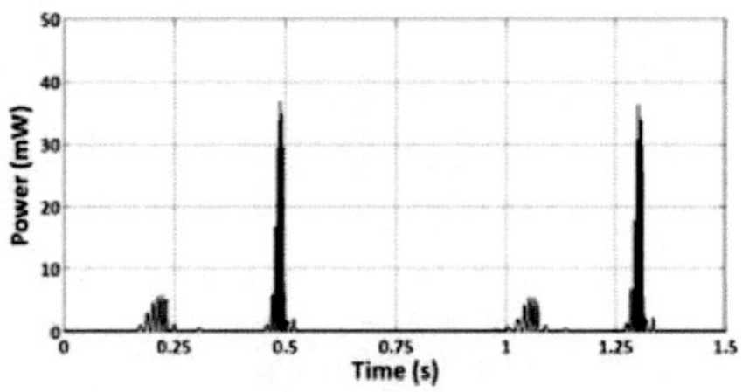

(h) 10 km/h

Fig. A.5 Voltage (on the left side) and power output (on the right side) data obtained experimentally for HAC3 (one magnet stack) and test person "2". The load resistance of 33 Ω is equal to the total coil resistance and identical across all measurements. Figures from [1]

A.1.6 HAC5 for Test Person "1"

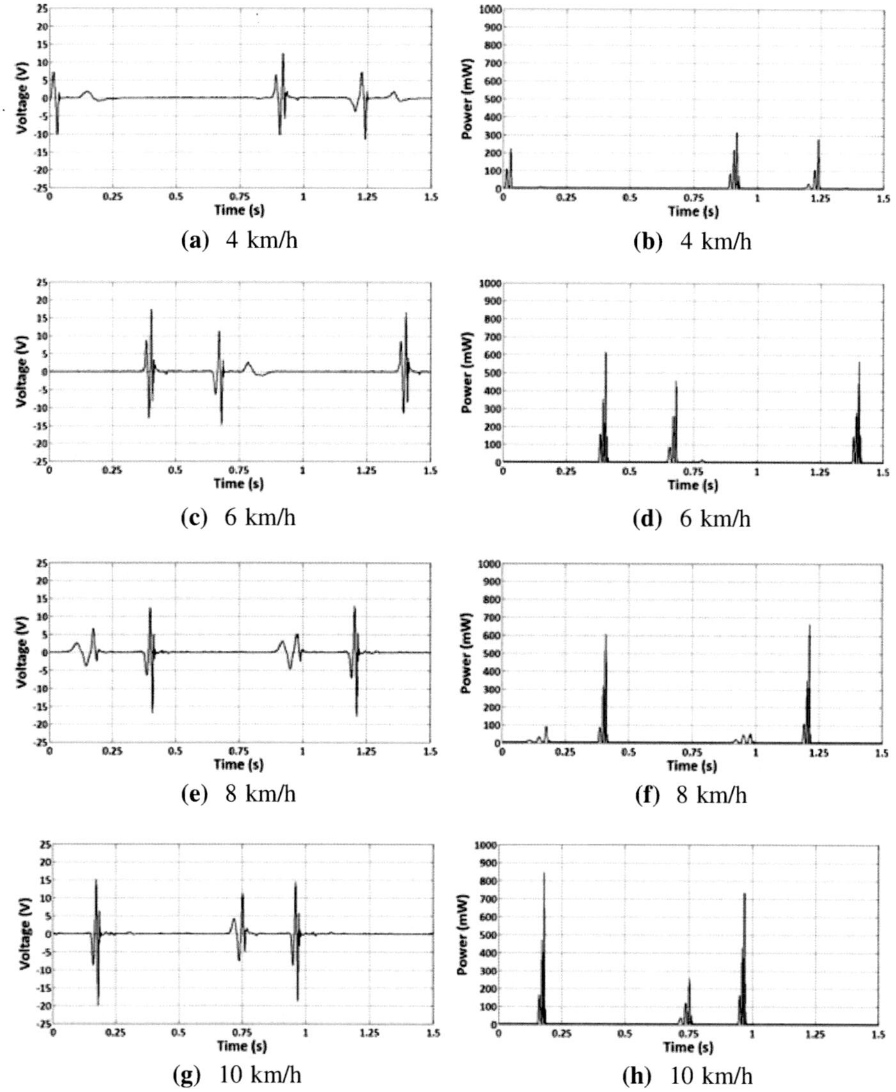

Fig. A.6 Voltage (on the left side) and power output (on the right side) data obtained experimentally for HAC5 and test person "1". The load resistance of 480 Ω is equal to the total coil resistance and identical across all measurements. Figures from [1]

A.1.7 HAC5 for Test Person "2"

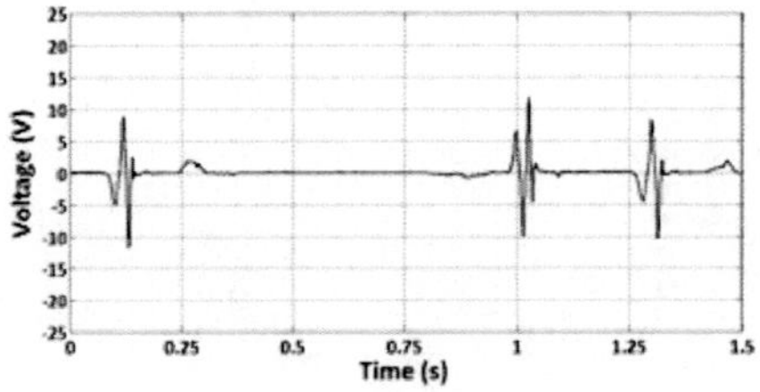

(a) 4 km/h

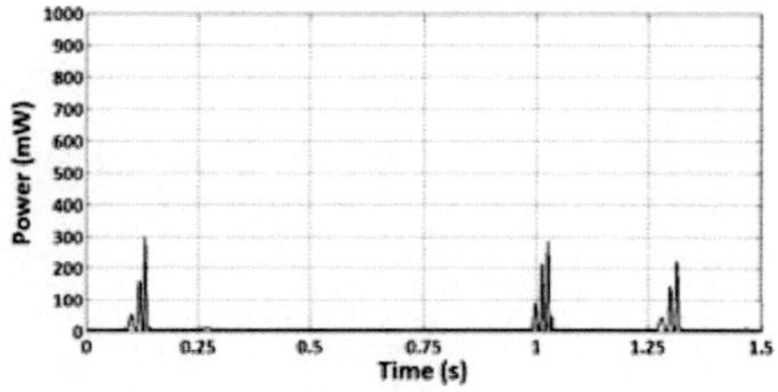

(b) 4 km/h

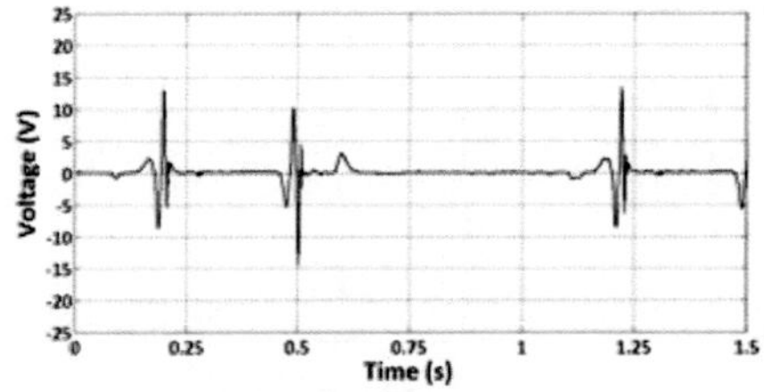

(c) 6 km/h

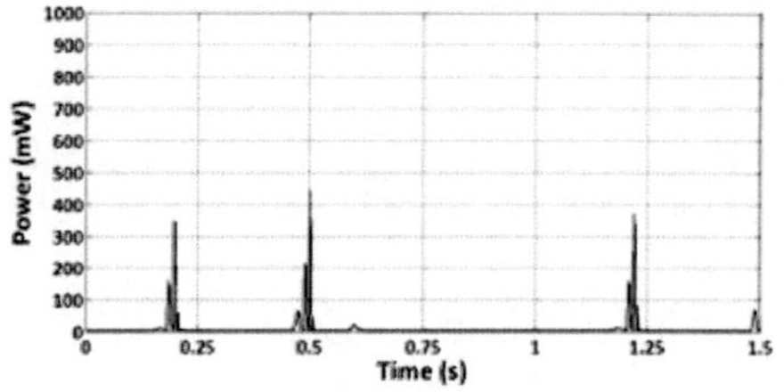

(d) 6 km/h

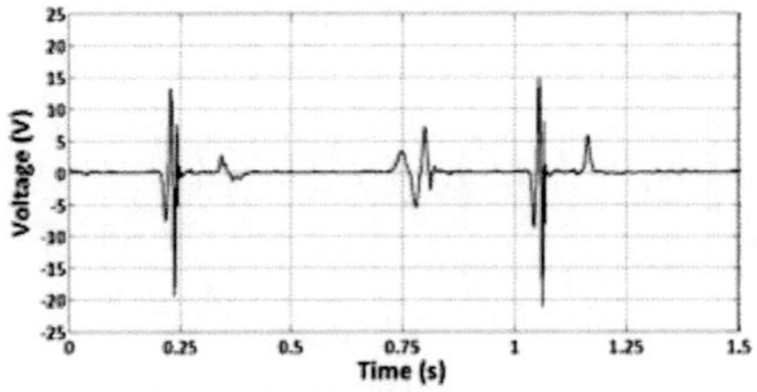

(e) 8 km/h

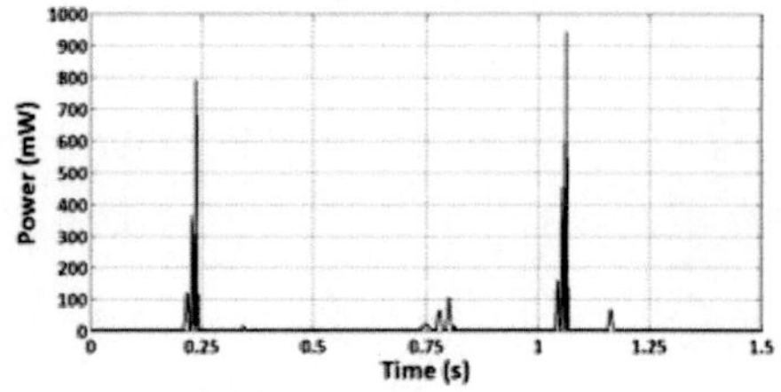

(f) 8 km/h

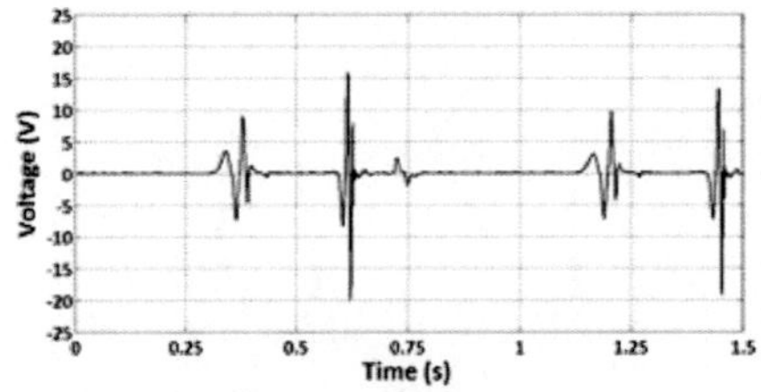

(g) 10 km/h

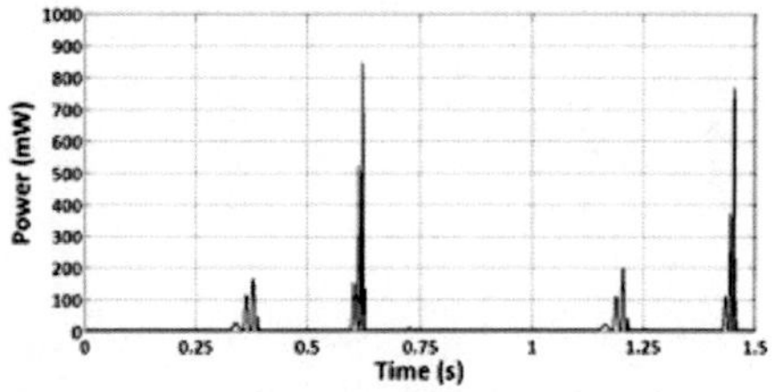

(h) 10 km/h

Fig. A.7 Voltage (on the left side) and power output (on the right side) data obtained experimentally for HAC5 and test person "2". The load resistance of 480 Ω is equal to the total coil resistance and identical across all measurements. Figures from [1]

A.1.8 HAC5 F for Test Person "1"

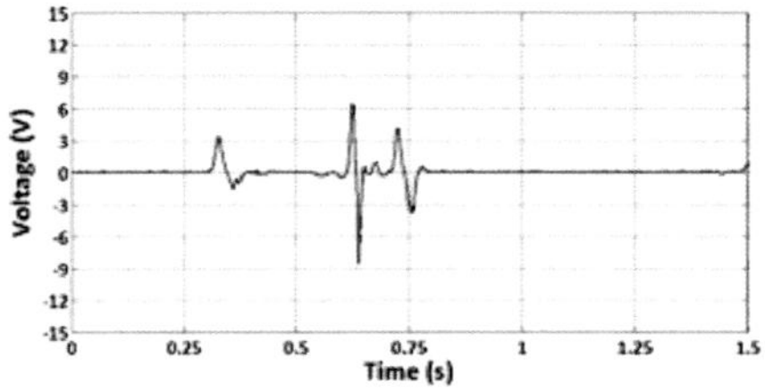

(a) 4 km/h

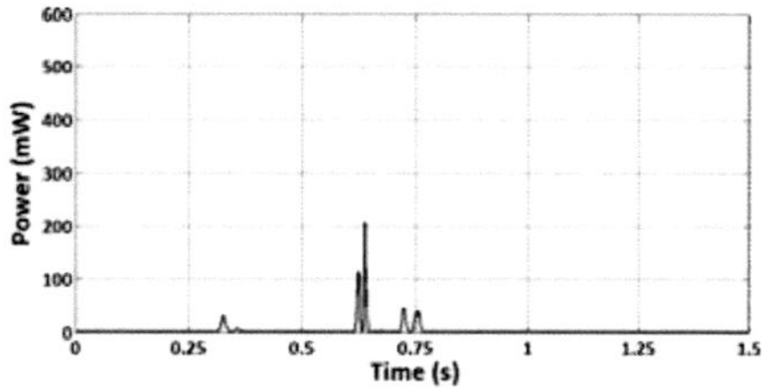

(b) 4 km/h

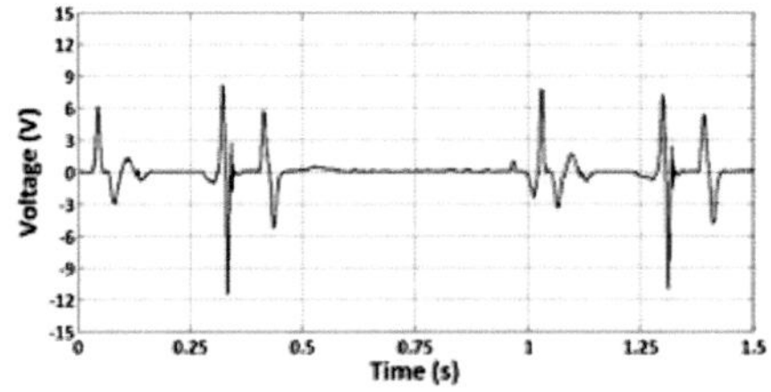

(c) 6 km/h

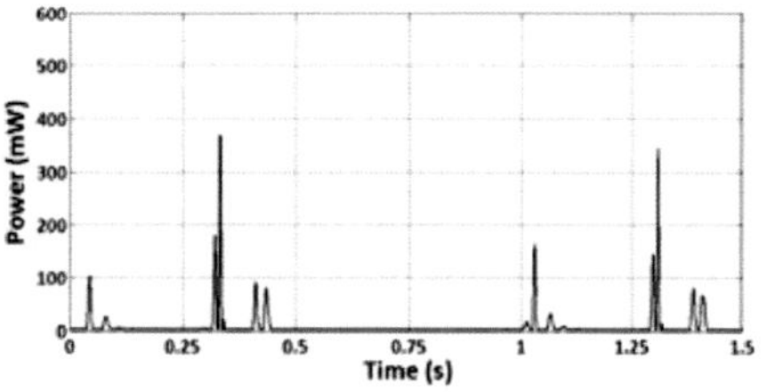

(d) 6 km/h

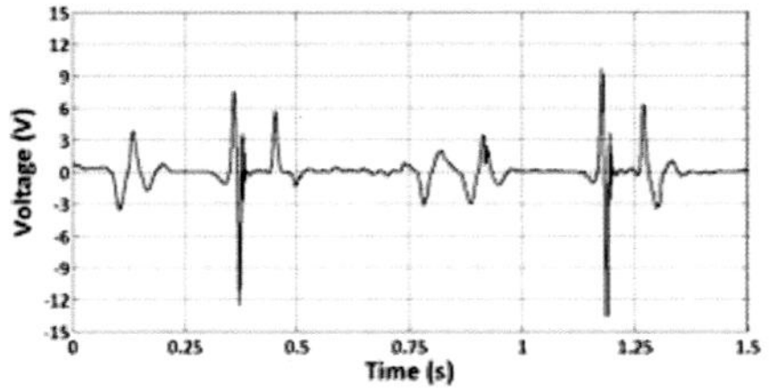

(e) 8 km/h

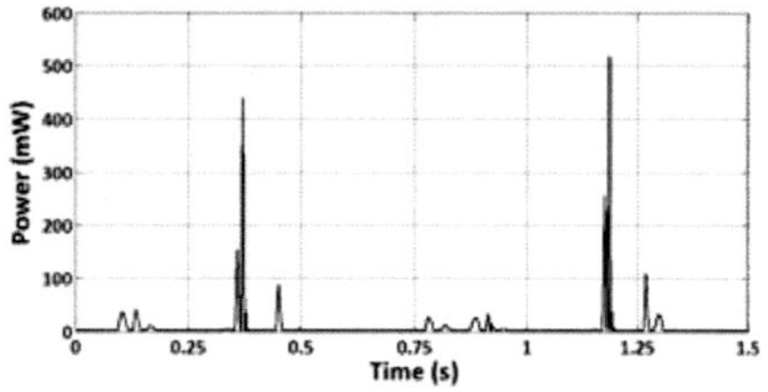

(f) 8 km/h

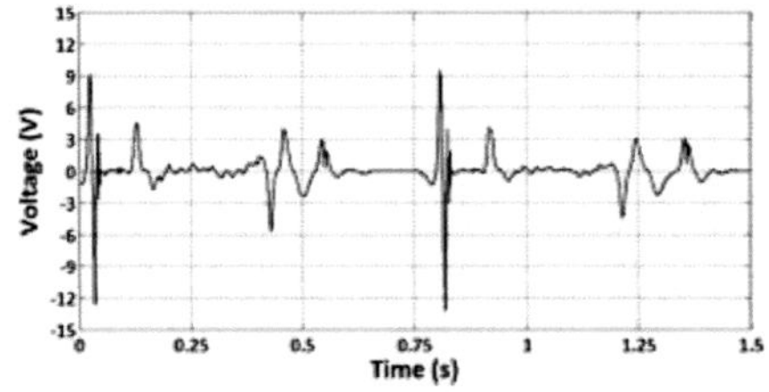

(g) 10 km/h

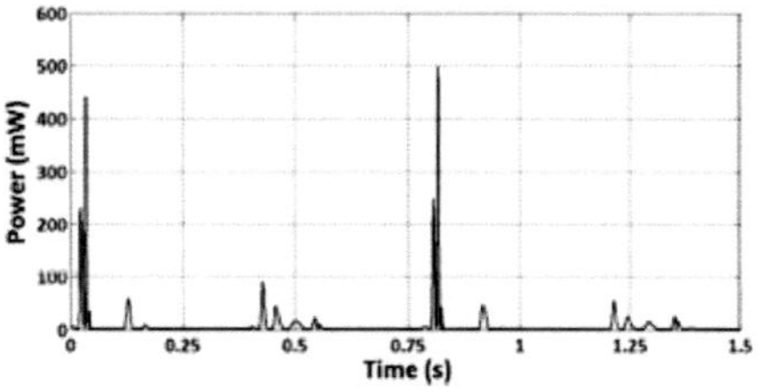

(h) 10 km/h

Fig. A.8 Voltage (on the left side) and power output (on the right side) data obtained experimentally for HAC5 F (spring-loaded version of HAC5) and test person "1". The load resistance of 360 Ω is equal to the total coil resistance and identical across all measurements. Figures from [1]

A.1.9 HAC5 F for Test Person "2"

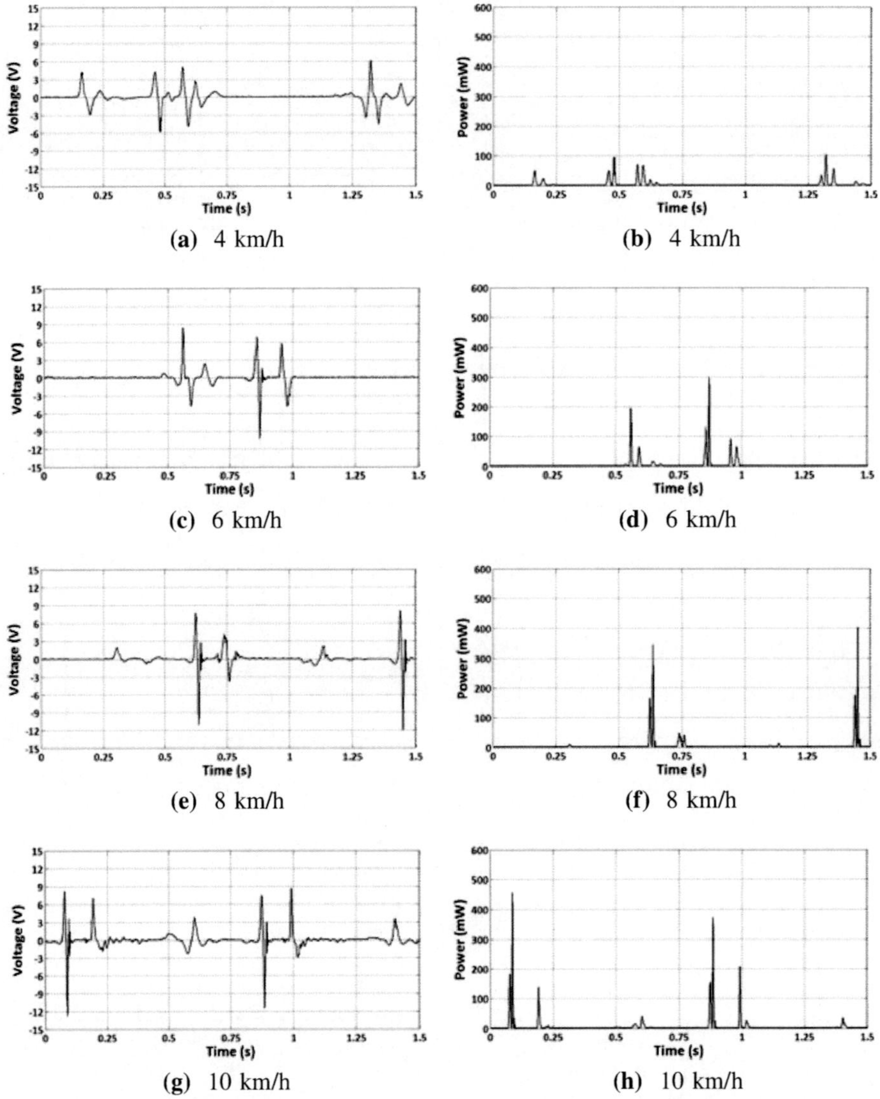

(a) 4 km/h **(b)** 4 km/h

(c) 6 km/h **(d)** 6 km/h

(e) 8 km/h **(f)** 8 km/h

(g) 10 km/h **(h)** 10 km/h

Fig. A.9 Voltage (on the left side) and power output (on the right side) data obtained experimentally for HAC5 F (spring-loaded version of HAC5) and test person "2". The load resistance of 360 Ω is equal to the total coil resistance and identical across all measurements. Figures from [1]

A.1.10 HAC6 for Test Person "1"

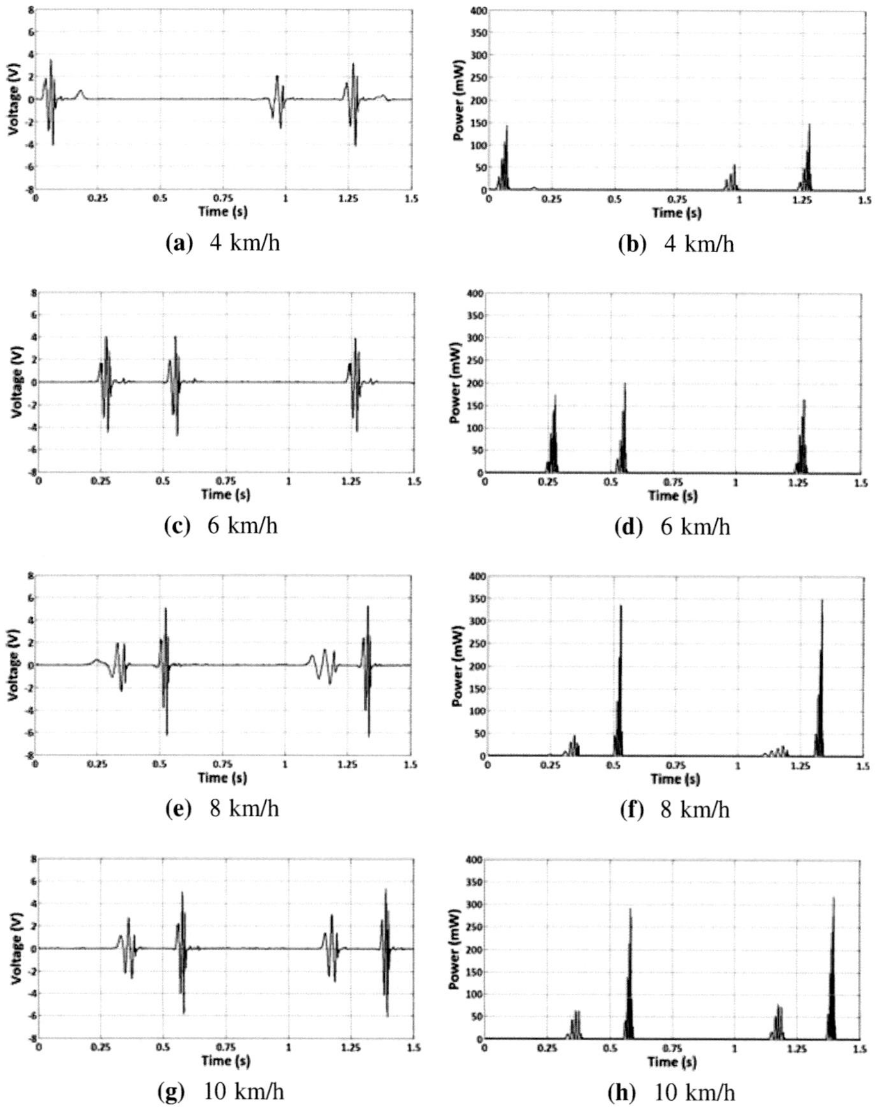

Fig. A.10 Voltage (on the left side) and power output (on the right side) data obtained experimentally for HAC6 and test person "1". The load resistance of 118 Ω is equal to the total coil resistance and identical across all measurements. Figures from [1]

A.1.11 HAC6 for Test Person "2"

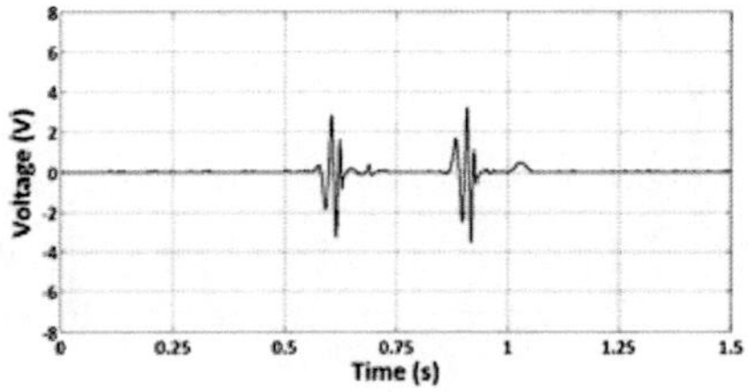

(a) 4 km/h

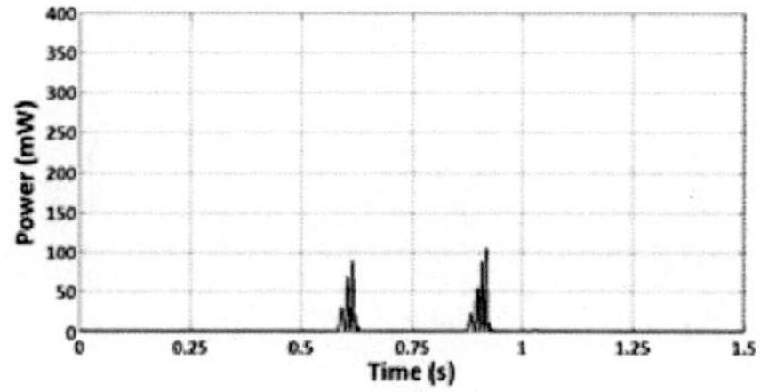

(b) 4 km/h

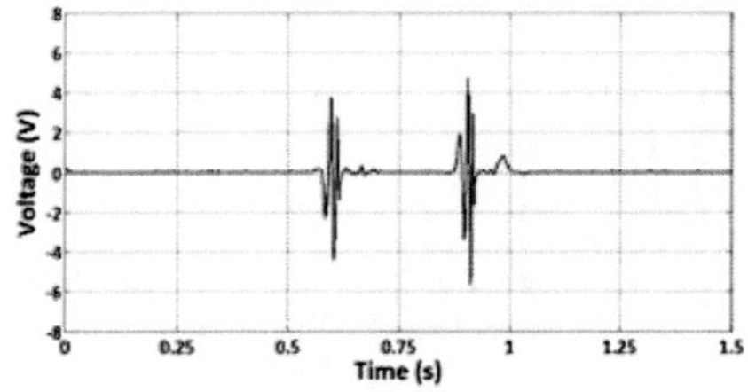

(c) 6 km/h

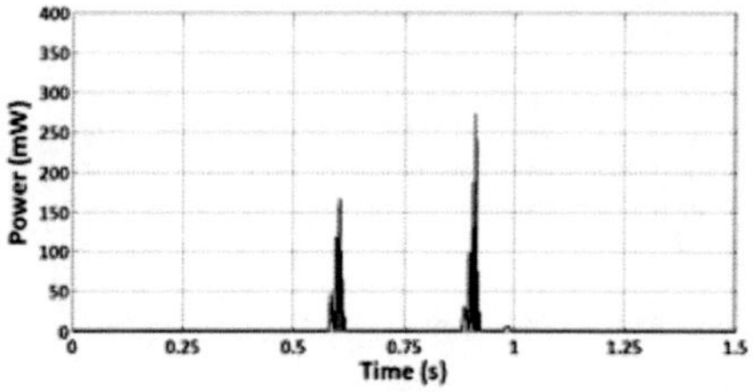

(d) 6 km/h

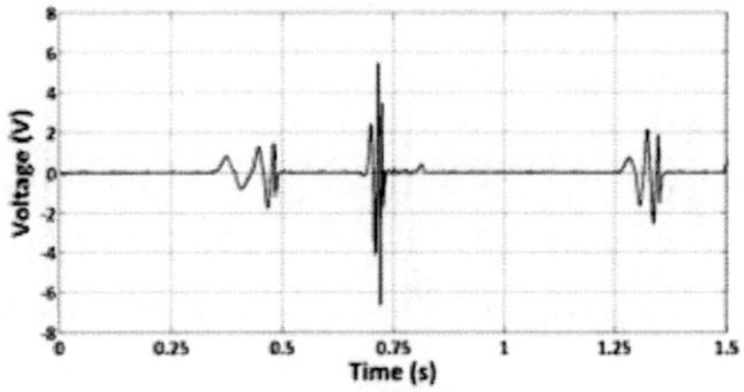

(e) 8 km/h

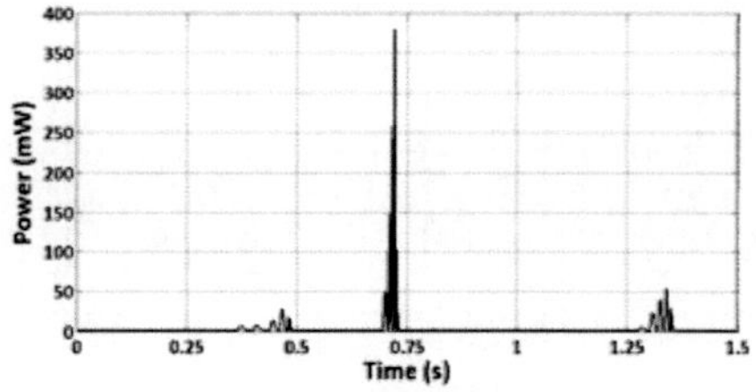

(f) 8 km/h

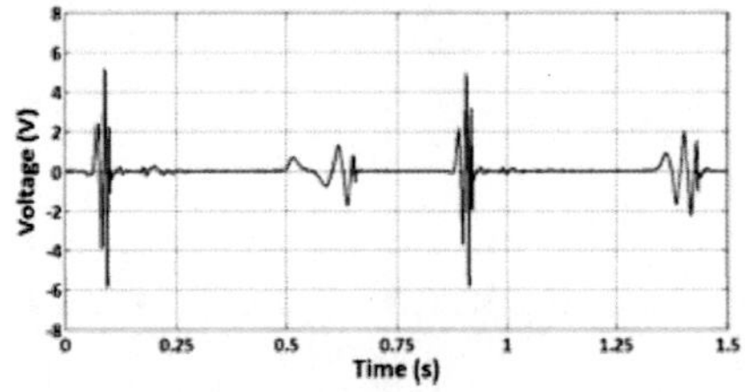

(g) 10 km/h

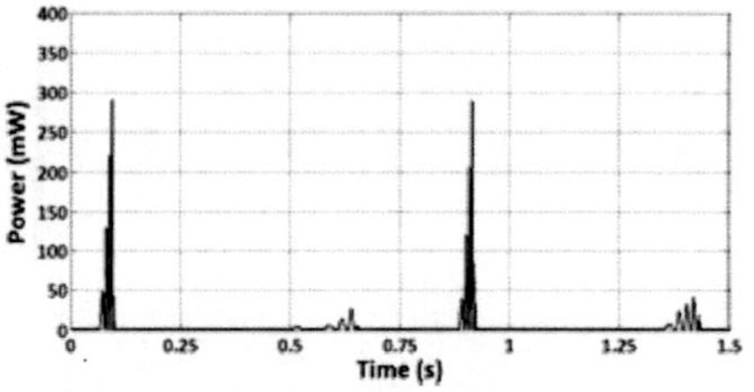

(h) 10 km/h

Fig. A.11 Voltage (on the left side) and power output (on the right side) data obtained experimentally for HAC6 and test person "2". The load resistance of 118 Ω is equal to the total coil resistance and identical across all measurements. Figures from [1]

A.2 Second Optimization Run—Scatter Graphs

In the following, a number of scatter graphs are given, which complement the results of the second optimization run performed in Chap. 5. The first scatter graph in Fig. A.12 presents results for HAC1. It shows the designs found by the optimization algorithm plotten against the coil length (named l_{Spule} in the graphs) and the wall thickness (named d_s in the graphs) of the cylindrical coils. As can be noticed, neither of the two parameters leads to a large power output, because the majority of designs is gathered at the bottom left of the graph at coil lengths of roughly 6 mm and wall thicknesses of 1 mm. The tendency towards a small wall thickness means, that the magnet diameter can be comparatively large within the same building space.

In the case of HAC3 (Fig. A.13), again the trend goes toward a relatively small coil length with most designs gathering at a length of approximately 4 mm. The wall thickness shows a larger variation than for HAC1 with the majority of designs clustered between 2.2 and 2.8 mm. In this architecture, the coils lie within the moving ring magnets. Consequently, the coil wall thickness cannot be small, otherwise both the number of windings and the coils cross-sectional area that is penetrated by magnetic flux lines and contributes to the induced voltage, would be small and subsequently lead to small power output.

For the rectangular architecture HAC5, the scatter graph in Fig. A.14 plots the designs found by the optimization algorithm against the height (named l_{Spule} in the graph) and width (named h_{coil} in the graph) of the coils. As far as the coil height is

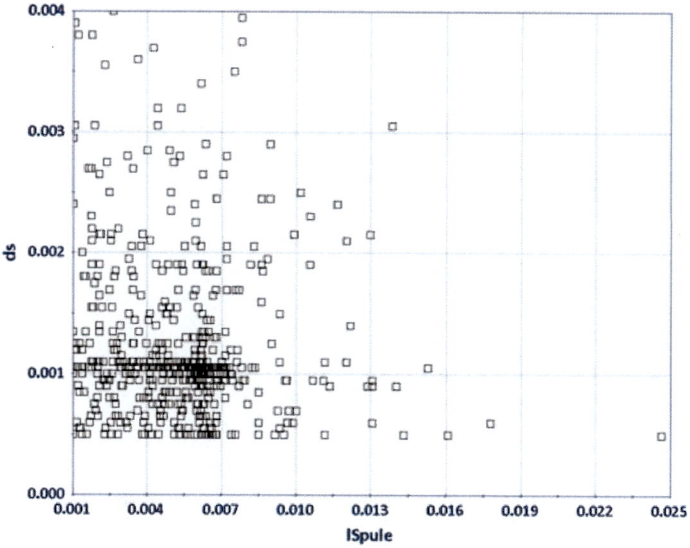

Fig. A.12 Scatter graph showing the distribution of the designs found for HAC1 during the optimization process along the length and wall thickness of the coils. Lengths scale in meters. Figure from [1]

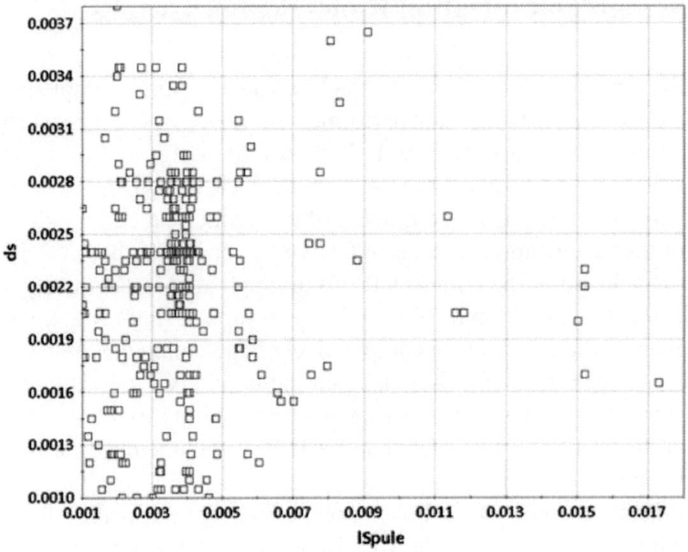

Fig. A.13 Scatter graph showing the distribution of the designs found for HAC3 during the optimization process along the length and wall thickness of the coils. Lengths scale in meters. Figure from [1]

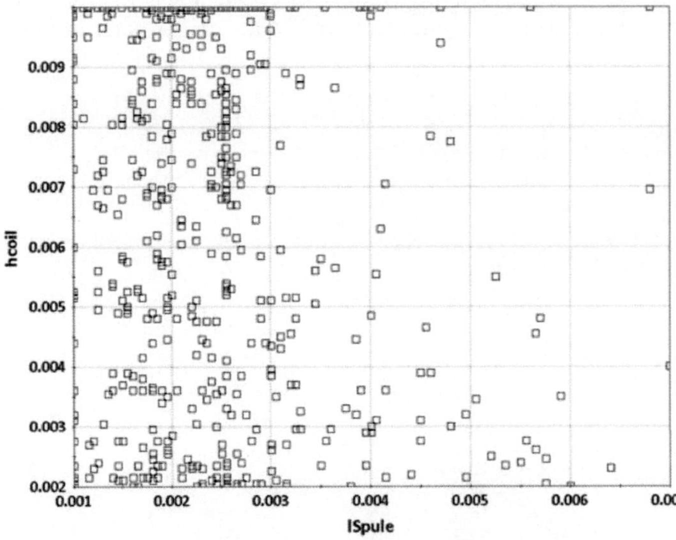

Fig. A.14 Scatter graph showing the distribution of the designs found for HAC5 during the optimization process along the height (l_{Spule}) and width (h_{coil}) of the coils. Lengths scale in meters. Figure from [1]

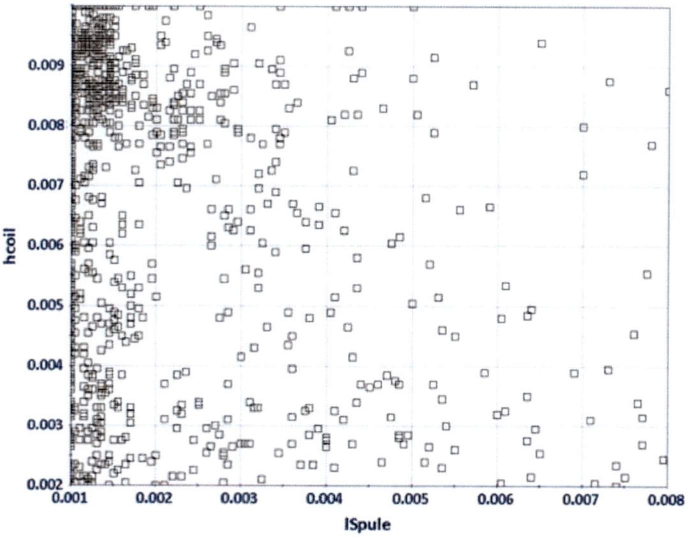

Fig. A.15 Scatter graph showing the distribution of the designs found for HAC6 during the optimization process along the height (l_{Spule}) and width (h_{coil}) of the coils. Lengths scale in meters. Figure from [1]

concerned, the majority of designs is gathered below heights of 3 mm. In terms of the coil width, the designs are spread more widely. However, a slight tendency towards the maximum coil width of 10 mm (which is also the parameter value for the best design of the optimization as presented in Chap. 5) can be noticed.

For the double Halbach array (HAC6), a large number of designs are clustered at small coil heights of 1.5 mm or less and large coil widths of 8 to 9 mm as depicted in Fig. A.15.

Reference

1. K. Ylli, Energy harvesting from the swing motion of the human leg (2019)

Index

Note: Page numbers followed by f, t, and equ indicates figures, tables and equations respectively

K. Ylli and Y. Manoli, *Energy Harvesting for Wearable Sensor Systems*, Springer Series in Advanced Microelectronics 62,
https://doi.org/10.1007/978-981-33-4448-8